A TEXTBOOK OF
APPLIED MICROBIOLOGY

A TEXTBOOK OF APPLIED MICROBIOLOGY

Dr. Sarfaraz Ahmad

ANMOL PUBLICATIONS PVT. LTD.

NEW DELHI-110 002 (INDIA)

ANMOL PUBLICATIONS PVT. LTD.
Regd. Office: 4360/4, Ansari Road, Daryaganj,
New Delhi-110002 (India)
Tel.: 23278000, 23261597, 23286875, 23255577
Fax: 91-11-23280289
Email: anmolpub@gmail.com
Visit us at: www.anmolpublications.com

Branch Office: No. 1015, Ist Main Road, BSK IIIrd Stage
IIIrd Phase, IIIrd Block, Bengaluru-560 085 (India)
Tel.: 080-41723429 • Fax: 080-26723604
Email: anmolpublicationsbangalore@gmail.com

A Textbook of Applied Microbiology

First Edition, 2011

ISBN 978-81-261-4731-1

PRINTED IN INDIA

Printed at Mehra Offset Press, Delhi.

Contents

Preface

The possibility that microorganisms exist was discussed for many centuries before their actual discovery in the 17th century. The existence of unseen microbiological life was postulated by Jainism which is based on Mahavira's teachings as early as 6th century BCE. Paul Dundas notes that Mahavira asserted existence of unseen microbiological creatures living in earth, water, air and fire. Jain scriptures also describe nigodas which are sub-microscopic creatures living in large clusters and having a very short life and are said to pervade each and every part of universe, even in tissues of plants and flesh of animals. Anton van Leeuwenhoek was one of the first people to observe microorganisms, using a microscope of his own design, and made one of the most important contributions to biology. Robert Hooke was the first to use a microscope to observe living things; his 1665 book *Micrographia* contained descriptions of plant cells.

Before Leeuwenhoek's discovery of microorganisms in 1675, it had been a mystery why grapes could be turned into wine, milk into cheese, or why food would spoil. Leeuwenhoek did not make the connection between these processes and microorganisms, but using a microscope, he did establish that there were forms of life that were not visible to the naked eye. Leeuwenhoek's discovery, along with subsequent observations by Lazzaro Spallanzani and Louis Pasteur, ended the long-held belief that life spontaneously appeared from non-living substances during the process of spoilage. Lazzaro Spallanzani found that boiling broth would sterilise it and kill any microorganisms in it. He also found that new microorganisms could only settle in a broth if the broth was exposed to the air. Louis Pasteur expanded upon Spallanzani's findings by exposing boiled broths to the air, in vessels that contained a filter to prevent all particles from passing through to the growth medium, and also in vessels with no filter at all, with air being admitted via a curved tube that would not allow dust particles to come in contact with the broth. By boiling the broth beforehand, Pasteur ensured that no microorganisms survived within the broths at the beginning of his experiment. Nothing grew in the broths in the course of Pasteur's experiment. This meant that the living organisms that grew in such broths came from outside, as spores on dust, rather than spontaneously generated within the broth. Thus, Pasteur dealt the death blow to the theory of spontaneous generation and supported germ theory.

Applied Microbiology is the study of microorganisms and their applications. A microorganism or microbe is an organism that is unicellular or lives in a colony of cellular organisms. The study of microorganisms is called microbiology, a subject that began with Anton van Leeuwenhoek's discovery of microorganisms in 1675, using a microscope of his own design. Microorganisms are very diverse; they include bacteria, fungi, archaea, and protists; microscopic plants (green algae); and animals such as plankton and the planarian. Some microbiologists also include viruses, but others consider these as non-living. Most microorganisms are unicellular (single-celled), but this is not universal, since some multicellular organisms are microscopic, while some unicellular protists and bacteria, like *Thiomargarita namibiensis*, are macroscopic and visible to the naked eye. Microorganisms live in all parts of the biosphere where there is liquid water, including soil, hot springs, on the ocean floor, high in the atmosphere and deep inside rocks within the Earth's crust. Microorganisms are critical to nutrient recycling in ecosystems as they act as decomposers. As some microorganisms can fix nitrogen, they are a vital part of the nitrogen cycle, and recent studies indicate that airborne microbes may play a role in precipitation and weather. Microbes are also exploited by people in biotechnology, both in traditional food and beverage preparation, and in modern technologies based on genetic engineering. However, pathogenic microbes are harmful, since they invade and grow within other organisms, causing diseases that kill people, other animals and plants.

The present book has been written in view of the students those who are related to this subject. Its increasing scope has really enhanced its significance and it appears not possible to present all of it in a single book. The book is comprehensive but not lengthy.

—*Author*

1

Introduction

Applied Microbiology is the study of microorganisms and their applications. A microorganism or microbe is an organism that is unicellular or lives in a colony of cellular organisms. The study of microorganisms is called microbiology, a subject that began with Anton van Leeuwenhoek's discovery of microorganisms in 1675, using a microscope of his own design.

Microorganisms are very diverse; they include bacteria, fungi, archaea, and protists; microscopic plants (green algae); and animals such as plankton and the planarian. Some microbiologists also include viruses, but others consider these as non-living. Most microorganisms are unicellular (single-celled), but this is not universal, since some multicellular organisms are microscopic, while some unicellular protists and bacteria, like *Thiomargarita namibiensis*, are macroscopic and visible to the naked eye. Microorganisms live in all parts of the biosphere where there is liquid water, including soil, hot springs, on the ocean floor, high in the atmosphere and deep inside rocks within the Earth's crust. Microorganisms are critical to nutrient recycling in ecosystems as they act as decomposers. As some microorganisms can fix nitrogen, they are a vital part of the nitrogen cycle, and recent studies indicate that airborne microbes may play a role in precipitation and weather. Microbes are also exploited by people in biotechnology, both in traditional food and beverage preparation, and in modern technologies based on genetic engineering. However, pathogenic microbes are harmful, since they invade and grow within other organisms, causing diseases that kill people, other animals and plants.

History

Evolution

Single-celled microorganisms were the first forms of life to develop on Earth, approximately 3–4 billion years ago. Further evolution was slow, and for about 3 billion years in the Precambrian eon, all organisms were microscopic. So, for most of the history of life on Earth the only forms of life were microorganisms. Bacteria, algae and fungi have been identified in amber that is 220 million years old, which shows that the morphology of microorganisms has changed little since the Triassic period.

Most microorganisms can reproduce rapidly and microbes such as bacteria can also freely exchange genes by conjugation, transformation and transduction between widely-divergent species. This horizontal gene transfer, coupled with a high mutation rate and many other means of genetic variation, allows microorganisms to swiftly evolve (via natural selection) to survive in new environments and respond to environmental stresses. This rapid evolution is important in medicine, as it has led to the recent development of 'super-bugs' — pathogenic bacteria that are resistant to modern antibiotics.

Pre-microbiology

The possibility that microorganisms exist was discussed for many centuries before their actual discovery in the 17th century. The existence of unseen microbiological life was postulated by Jainism which is based on Mahavira's teachings as early as 6th century BCE. Paul Dundas notes that Mahavira asserted existence of unseen microbiological creatures living in earth, water, air and fire. Jain scriptures also describe nigodas which are sub-microscopic creatures living in large clusters and having a very short life and are said to pervade each and every part of universe, even in tissues of plants and flesh of animals. However, the earliest known idea to indicate the possibility of diseases spreading by yet unseen organisms was that of the Roman scholar Marcus Terentius Varro in a 1st century BC book titled *On Agriculture* in which he warns against locating a homestead near swamps:

> *"...and because there are bred certain minute creatures which cannot be seen by the eyes, which float in the air and enter the body through the mouth and nose and there cause serious diseases."*

In *The Canon of Medicine* (1020), Abu Ali ibn Sinâ (Avicenna) stated that bodily secretion is contaminated by foul foreign earthly bodies before being infected. He also hypothesized that tuberculosis and other diseases might be contagious, *i.e.* that they were infectious diseases, and used quarantine to limit their spread.

When the Black Death bubonic plague reached Andalusia in Spain, in the 14th century, Ibn Khatima wrote that infectious diseases were caused by contagious "minute bodies" that enter the human body. Later, in 1546, Girolamo Fracastoro proposed that epidemic diseases were caused by transferable seedlike entities that could transmit infection by direct or indirect contact, or even without contact over long distances.

All these early claims about the existence of microorganisms were speculative and were not based on any data or science. Microorganisms were neither proven, observed, nor correctly and accurately described until the 17th century. The reason for this was that all these early studies lacked the microscope.

History of Microorganisms' Discovery

Anton van Leeuwenhoek was one of the first people to observe microorganisms, using a microscope of his own design, and made one of the most important contributions to biology. Robert Hooke was the first to use a microscope to observe living things; his 1665 book *Micrographia* contained descriptions of plant cells.

Before Leeuwenhoek's discovery of microorganisms in 1675, it had been a mystery why grapes could be turned into wine, milk into cheese, or why food would spoil. Leeuwenhoek did not make the connection between these processes and microorganisms, but using a microscope, he did establish that there were forms of life that were not visible to the naked eye. Leeuwenhoek's discovery, along with subsequent observations by Lazzaro Spallanzani and Louis Pasteur, ended the long-held belief that life spontaneously appeared from non-living substances during the process of spoilage.

Lazzaro Spallanzani found that boiling broth would sterilise it and kill any microorganisms in it. He also found that new microorganisms could only settle in a broth if the broth was exposed to the air. Louis Pasteur expanded upon Spallanzani's findings by exposing boiled broths to the air, in vessels that contained a filter to prevent all particles from passing through to the growth medium, and also in vessels with no filter at all, with air being admitted via a curved tube that would not allow dust particles to come in contact with the broth. By boiling the broth beforehand, Pasteur ensured that no microorganisms survived within the broths at the beginning of his experiment. Nothing grew in the broths in the course of Pasteur's experiment. This meant that the living organisms that grew in such broths came from outside, as spores on dust, rather than spontaneously generated within the broth. Thus, Pasteur dealt the death blow to the theory of spontaneous generation and supported germ theory.

In 1876, Robert Koch established that microbes can cause disease. He found that the blood of cattle who were infected with anthrax always had large numbers of *Bacillus anthracis*. Koch found that he could transmit anthrax from one animal to another by taking a small sample of blood from the infected animal and injecting it into a healthy one, and this caused the healthy animal to become sick. He also found that he could grow the bacteria in a nutrient broth, then inject it into a healthy animal, and cause illness. Based on these experiments, he devised criteria for establishing a causal link between a microbe and a disease and these are now known as Koch's postulates. Although these postulates cannot be applied in all cases, they do retain historical importance to the development of scientific thought and are still being used today.

Classification and Structure

Microorganisms can be found almost anywhere in the taxonomic organization of life on the planet. Bacteria and archaea are almost always microscopic, while a number of eukaryotes are also microscopic, including most protists, some fungi, as well as some animals and plants. Viruses are generally regarded as not living and therefore are not microbes, although the field of microbiology also encompasses the study of viruses.

Prokaryotes

Prokaryotes are organisms that lack a cell nucleus and the other membrane bound organelles. They are almost always unicellular, although some species such as myxobacteria can aggregate into complex structures as part of their life cycle. Consisting of two domains, bacteria and archaea, the prokaryotes are the most diverse and abundant group of organisms on Earth and inhabit practically all environments where some liquid water is available and the temperature is below +140 °C. They are found in sea water, soil, air, animals' gastrointestinal tracts, hot springs and even deep beneath the Earth's crust in rocks. Practically all surfaces which have not been specially sterilized are covered by prokaryotes. The number of prokaryotes on Earth is estimated to be around five million trillion trillion, or 5×10^{30}, accounting for at least half the biomass on Earth.

Bacteria

Bacteria are practically all invisible to the naked eye, with a few extremely rare exceptions, such as *Thiomargarita namibiensis*. They lack membrane-bound organelles, and can function and reproduce as individual cells, but often aggregate in multicellular colonies. Their genome is usually a single loop of DNA, although they can also harbor small pieces of DNA called plasmids. These plasmids can be transferred between cells through bacterial conjugation. Bacteria are surrounded by a cell wall, which provides strength and rigidity to their cells. They reproduce by binary fission or sometimes by budding, but do not undergo sexual reproduction. Some species form extraordinarily resilient spores, but for bacteria this is a mechanism for survival, not reproduction. Under optimal conditions bacteria can grow extremely rapidly and can double as quickly as every 10 minutes.

Archaea

Archaea are also single-celled organisms that lack nuclei. In the past, the differences between bacteria and archaea were not recognised and archaea were classified with bacteria as part of the kingdom Monera. However, in 1990 the microbiologist Carl Woese proposed the three-domain system that divided living things into bacteria, archaea and eukaryotes. Archaea differ from bacteria in both their genetics and biochemistry. For example, while bacterial cell membranes are made from phosphoglycerides with ester bonds, archaean membranes are made of ether lipids.

Archaea were originally described in extreme environments, such as hot springs, but have since been found in all types of habitats. Only now are scientists beginning to realize how common archaea are in the environment, with crenarchaeota being the most common form of life in the ocean, dominating ecosystems below 150 m in depth. These organisms are also common in soil and play a vital role in ammonia oxidation.

Eukaryotes

Most living things which are visible to the naked eye in their adult form are eukaryotes, including humans. However, a large number of eukaryotes are also microorganisms. Unlike bacteria and archaea, eukaryotes contain organelles such as the cell nucleus, the Golgi apparatus and mitochondria in their cells. The nucleus is an organelle which houses the DNA that makes up a cell's genome. DNA itself is arranged in complex chromosomes. Mitochondria are organelles vital in metabolism as they are the site of the citric acid cycle and oxidative phosphorylation. They evolved from symbiotic bacteria and retain a remnant genome. Like bacteria, plant cells have cell walls, and contain organelles such as chloroplasts in addition to the organelles in other eukaryotes. Chloroplasts produce energy from light by photosynthesis, and were also originally symbiotic bacteria.

Unicellular eukaryotes are those eukaryotic organisms that consist of a single cell throughout their life cycle. This qualification is significant since most multicellular eukaryotes consist of a single cell called a zygote at the beginning of their life cycles. Microbial eukaryotes can be either haploid or diploid, and some organisms have multiple cell nuclei. However, not all microorganisms are unicellular as some microscopic eukaryotes are made from multiple cells.

Protists

Of eukaryotic groups, the protists are most commonly unicellular and microscopic. This is a highly diverse group of organisms that are not easy to classify. Several algae species are multicellular protists, and slime molds have unique life cycles that involve switching between unicellular, colonial, and multicellular forms. The number of species of protozoa is uncertain, since we may have identified only a small proportion of the diversity in this group of organisms.

Animals

Mostly animals are multicellular, but some are too small to be seen by the naked eye. Microscopic arthropods include dust mites and spider mites. Microscopic crustaceans include copepods and the cladocera, while many nematodes are too small to be seen with the naked eye. Another particularly common group of microscopic animals are the rotifers, which are filter feeders that are usually found in fresh water. Micro-animals reproduce both sexually and asexually and may reach new habitats as eggs that survive harsh environments that would kill the adult animal. However, some simple animals, such as rotifers and nematodes, can dry out completely and remain dormant for long periods of time.

Fungi

The fungi have several unicellular species, such as baker's yeast (*Saccharomyces cerevisiae*) and fission yeast (*Schizosaccharomyces pombe*). Some fungi, such as the pathogenic yeast *Candida albicans*, can undergo phenotypic switching and grow as single cells in some environments, and filamentous hyphae in others. Fungi reproduce both asexually, by budding or binary fission, as well by producing spores, which are called conidia when produced asexually, or basidiospores when produced sexually.

Plants

The green algae are a large group of photosynthetic eukaryotes that include many microscopic organisms. Although some green algae are classified as protists, others such as charophyta are classified with embryophyte plants, which are the most familiar group of land plants. Algae can grow as single cells, or in long chains of cells. The green algae include unicellular and colonial flagellates, usually but not always with two flagella per cell, as well as various colonial, coccoid, and filamentous forms. In the Charales, which are the algae most closely related to higher plants, cells differentiate into several distinct tissues within the organism. There are about 6000 species of green algae.

Habitats and Ecology

Microorganisms are found in almost every habitat present in nature. Even in hostile environments such as the poles, deserts, geysers, rocks, and the deep sea. Some types of microorganisms have adapted to the extreme conditions and sustained colonies; these organisms are known as extremophiles. Extremophiles have been isolated from rocks as much as 7 kilometres below the Earth's surface, and it has been suggested that the amount of living organisms below the Earth's surface may be comparable with the amount of life on or above the surface. Extremophiles have been known to survive for a prolonged time in a vacuum, and can be highly resistant to radiation, which may even allow them to survive in space. Many types of microorganisms have intimate symbiotic relationships with other larger organisms; some of which are mutually beneficial (mutualism), while others can be damaging to the host organism (parasitism). If microorganisms can cause disease in a host they are known as pathogens.

Extremophiles

Extremophiles are microorganisms which have adapted so that they can survive and even thrive in conditions that are normally fatal to most life-forms. For example, some species have been found in the following extreme environments:

- Temperature: as high as 130 °C (266 °F), as low as –17 °C (1.4 °F)
- Acidity/alkalinity: less than pH 0, up to pH 11.5
- Salinity: up to saturation
- Pressure: up to 1,000-2,000 atm, down to 0 atm (e.g. vacuum of space)
- Radiation: up to 5kGy.

Extremophiles are significant in different ways. They extend terrestrial life into much of the Earth's hydrosphere, crust and atmosphere, their specific evolutionary adaptation mechanisms to their extreme environment can be exploited in bio-technology, and their very existence under such extreme conditions increases the potential for extraterrestrial life.

Soil Microbes

The nitrogen cycle in soils depends on the fixation of atmospheric nitrogen. One way this can occur is in the nodules in the roots of legumes that contain symbiotic bacteria of the genera *Rhizobium*, *Mesorhizobium*, *Sinorhizobium*, *Bradyrhizobium*, and *Azorhizobium*.

Symbiotic Microbes

Symbiotic microbes such as fungi and algae form an association in lichen. Certain fungi form mycorrhizal symbioses with trees that increase the supply of nutrients to the tree.

Importance

Microorganisms are vital to humans and the environment, as they participate in the Earth's element cycles such as the carbon cycle and nitrogen cycle, as well as fulfilling other vital roles in virtually all ecosystems, such as recycling other organisms' dead remains and waste products through decomposition. Microbes also have an important place in most higher-order multicellular organisms as symbionts. Many blame the failure of Biosphere 2 on an improper balance of microbes.

Use in Food

Microorganisms are used in brewing, winemaking, baking, pickling and other food-making processes.

They are also used to control the fermentation process in the production of cultured dairy products such as yogurt and cheese. The cultures also provide flavour and aroma, and inhibit undesirable organisms.

Use in Water Treatment

Specially-cultured microbes are used in the biological treatment of sewage and industrial waste effluent, a process known as bioaugmentation.

Use in Energy

Microbes are used in fermentation to produce ethanol, and in biogas reactors to produce methane. Scientists are researching the use of algae to produce liquid fuels, and bacteria to convert various forms of agricultural and urban waste into usable fuels.

Use in Science

Microbes are also essential tools in biotechnology, biochemistry, genetics, and molecular biology. The yeasts (*Saccharomyces cerevisiae*) and fission yeast (*Schizosaccharomyces pombe*) are important model organisms in science, since they are simple eukaryotes that can be grown rapidly in large numbers and are easily manipulated. They are particularly valuable in genetics, genomics and proteomics. Microbes can be harnessed for uses such as creating steroids and treating skin diseases. Scientists are also considering using microbes for living fuel cells, and as a solution for pollution.

Use in Warfare

In the Middle Ages, diseased corpses were thrown into castles during sieges using catapults or other siege engines. Individuals near the corpses were exposed to the deadly pathogen and were likely to spread that pathogen to others.

Importance in Human Health

Human Digestion

Microorganisms can form an endosymbiotic relationship with other, larger organisms. For example, the bacteria that live within the human digestive system contribute to gut immunity, synthesise vitamins such as folic acid and biotin, and ferment complex indigestible carbohydrates.

Diseases and Immunology

Microorganisms are the cause of many infectious diseases. The organisms involved include pathogenic bacteria, causing diseases such as plague, tuberculosis and anthrax; protozoa, causing diseases such as malaria, sleeping sickness and toxoplasmosis; and also fungi causing diseases such as ringworm, candidiasis or histoplasmosis. However, other diseases such as influenza, yellow fever or AIDS are caused by pathogenic viruses, which are not usually classified as living organisms and are not therefore microorganisms by the strict definition. As of 2007, no clear examples of archaean pathogens are known, although a relationship has been proposed between the presence of some methanogens and human periodontal disease.

Importance in Ecology

Microbes are critical to the processes of decomposition required to cycle nitrogen and other elements back to the natural world.

Hygiene

Hygiene is the avoidance of infection or food spoiling by eliminating microorganisms from the surroundings. As microorganisms, particularly bacteria, are found practically everywhere, this means in most cases the reduction of harmful microorganisms to acceptable levels. However, in some cases it is required that an object or substance be completely sterile, i.e. devoid of all living entities and viruses. A good example of this is a hypodermic needle.

In food preparation microorganisms are reduced by preservation methods (such as the addition of vinegar), clean utensils used in preparation, short storage periods or by cool temperatures. If complete sterility is needed, the two most common methods are irradiation and the use of an autoclave, which resembles a pressure cooker.

There are several methods for investigating the level of hygiene in a sample of food, drinking water, equipment etc. Water samples can be filtrated through an extremely fine filter. This filter is then placed in a nutrient medium. Microorganisms on the filter then grow to form a visible colony. Harmful microorganisms can be detected in food by placing a sample in a nutrient broth designed to enrich the organisms in question. Various methods, such as selective media or PCR, can then be used for detection. The hygiene of hard surfaces, such as cooking pots, can be tested by touching them with a solid piece of nutrient medium and then allowing the microorganisms to grow on it.

There are no conditions where all microorganisms would grow, and therefore often several different methods are needed. For example, a food sample might be analysed on three different nutrient mediums designed to indicate the presence of "total" bacteria (conditions where many, but not all, bacteria grow), molds (conditions where the growth of bacteria is prevented by e.g. antibiotics) and coliform bacteria (these indicate a sewage contamination).

Microbial Technologies Commercial Utilization

Microorganisms are capable of growing on a wide range of substrates and can produce a remarkable spectrum of products. The relatively recent advent of *in vitro* genetic manipulation has extended the range of products that may be produced by microorganisms and has provided new methods for increasing the yields of existing ones. The commercial exploitation of the biochemical diversity of microorganisms has resulted in the development of the fermentation industry and the techniques of genetic manipulation have given this well-established industry the opportunity to develop new processes and to improve existing ones.

The term fermentation is derived from the Latin verb *fervere,* to boil, which describes the appearance of the action of yeast on extracts of fruitor malted grain during the production of alcoholic beverages. However, fermentation is interpreted differently by microbiologists and biochemists. To a microbiologist the word means any process for the production of a product by the mass culture of microorganisms. To a biochemist, however, the word means an energy-generating process in which organic compounds act as both electron donors and acceptors, that is, an anaerobic process where energy is produced without the participation of oxygen or other inorganic electron acceptors.

Microbial Growth

The growth of a microorganism may result in the production of a range of metabolites but to produce a particular metabolite the desired organism must be grown under precise cultural conditions at a particular growth rate. If a microorganism is introduced into a nutrient medium that supports its growth, the inoculated culture will pass through a number of stages and the system is termed batch culture. Initially, growth does not occur and this period is referred to as the lag phase and may be considered a period of adaptation.

As already stated, the cessation of growth in a batch culture may be due to the exhaustion of a nutrient component or the accumulation of a toxic product. However, provided that the growth medium is designed such that growth is limited by the availability of a medium component, growth may be extended by addition of an aliquot of fresh medium to the vessel. If the fresh medium is added continuously, at an appropriate rate, and the culture vessel is fitted with an overflow device, such that culture is displaced by the incoming fresh medium, a continuous culture may be established. The growth of the cells in a continuous culture of this type is controlled by the availability of the growth limiting chemical component of the medium and, thus, the system is described as a chemostat. In'this system a steady-state is eventually achieved and the loss of biomass via the overflow is replaced by cell growth. The flow of medium through the system is described by the term dilution rate, *D*, which is equal to the rate of addition of medium divided by the working volume of the culture vessel.

Under steady-state conditions:

$dx/dt = 0$ and, therefore, $px = Dx$ and $p = D$.

Hence, the growth rate of the organisms is controlled by the dilution rate, which is an experimental variable. It will be recalled that under batch culture conditions an organism will grow at its maximum specific growth rate and, therefore, it is obvious that a continuous culture may be operated only at dilution rates below the maximum specific growth rate.

Thus, within certain limits, the dilution rate may be used to control the growth rate of a chemostat culture.

Applications of Fermentation

Microbial fermentations may be classified into the following major group:

(i) Those that produce microbial cells (biomass) as the product.

(ii) Those that produce microbial metabolites.

(iii) Those that produce microbial enzymes.

(iv) Those that modify a compound which is added to the fermentation-the transformation processes.

(v) Those that produce recombinant products.

Microbial Biomass

Microbial biomass is produced commercially as single cell protein (SCP) for human food or animal feed and as viable yeast cells to be used in the baking industry. The industrial production of bakers' yeast started in the early 1960s and yeast biomass was used as human food in Germany during the First World War. However, the development of large-scale processes for the production of microbial biomass as a source of commercial protein began in earnest in the late 1960s Several of the processes investigated did not come to fruition owing to political and economic problems but the establishment of the ICI Pruteen process for the production of bacterial SCP for animal feed was a milestone in the development of the fermentation industry. This process utilized continuous culture on an enormous scale (1 500 m3) and is an excellent example of the application of good engineering to the design of a microbiological process. However, the economics of the production of SCP as animal feed were marginal, which eventually led to the discontinuation of the Pruteen process. The technical expertise gained from the Pruteen process assisted ICI in collaborating with Rank Hovis MacDougall on a process for the production of fungal biomass to be used as human food? A continuous fermentation process for the production of *Fusarium graminearum* biomass was developed utilizing a 40 m3 air-lift fermenter. This process was based on sound economics and has proved to be a major economic success.

Microbial Metabolites

The kinetic description of batch culture may be rather misleading when considering the product-forming capacity of the culture during the various phases, for, although the metabolism of stationary phase cells is considerably different from that of logarithmic ones, it is by no means stationary. Bu'Lock *et aL6* proposed a descriptive terminology of the behaviour of microbial cells which considered the type of metabolism rather than the kinetics of growth. The term 'trophophase' was suggested to describe the log or exponential phase of a culture during which the sole products of metabolism are either essential to growth, such as amino acids, nucleotides, proteins, nucleic acids, lipids, carbohydrates, *etc.* or are the by-products of energy-yielding metabolism such as ethanol, acetone and butanol. The metabolites produced during the trophophase are referred to as primary metabolites.

At first sight it may seem anomalous that microorganisms produce compounds which do not appear to have any metabolic function and are certainly not by-products of catabolism as are, for example, ethanol and acetone. However, many secondary metabolites exhibit antimicrobial properties and, therefore, may be involved in competition in the natural environment; others have, since their discovery in idiophase cultures, been demonstrated to be produced during the trophophase where, it has been claimed, they act in some form of metabolic control.' Although the physiological role of secondary metabolism continues to be the subject of considerable debate its relevance to the fermentation industry is the commercial significance of the secondary metabolites.

The production of microbial metabolites may be achieved in continuous, as well as batch, systems. The chronological separation of trophophase and idiophase in batch culture may be studied in continuous culture in terms of dilution rate. Secondary metabolism will occur at relatively low dilution rates (growth rates) and, therefore, it should be remembered that secondary metabolism is a property of slow growing, as well as stationary, cells. The fact that secondary metabolites are produced by slow-growing organisms in continuous culture indicates that primary metabolism is continuing in idiophase-type cells. Thus, secondary metabolism is not switched on to remove an accumulation of metabolites synthesized entirely in a different phase; synthesis of the primary metabolic precursors continues through the period of secondary biosyn thesis.

The control of the onset of secondary metabolism has been studied extensively in batch culture and, to a lesser extent, in continuous culture. The outcome of this work is that a considerable amount of information is available on the interrelationships between the changes occurring in the medium and the cells at the onset of secondary metabolism and the control of the process. Primary metabolic precursors of secondary metabolites have been demonstrated to induce secondary metabolism, for example, tryptophan in alkaloid' biosynthesis and methionine in cephalosporin biosynthesis. On the other hand, medium components have been demonstrated to repress secondary metabolism, the earliest observation being that of Saltero and Johnson" in 1953 of the repressing effect of glucose on benzyl penicillin formation. Carbon sources that support high growth rates tend to support poor secondary metabolism.

Microbial Enzymes

The major commercial utilization of microbial enzymes is in the food and beverage industries although enzymes do have considerable application in clinical and analytical situations, as well as their use in washing powders. Most enzymes are synthesized in the logarithmic phase of batch culture and may, therefore, be considered as primary metabolites. However, some, for example the amylases of *Bacillus stearothermophilus,*" are produced by idiophase cultures and may be considered as equivalent to secondary metabolites. Enzymes may be produced from animals and plants as well as microbial sources but the production by microbial fermentation is the most economic and convenient method. Furthermore, it is now possible to engineer microbial cells to produce animal or plant enzymes.

Transformation Processes

As well as the use of microorganisms to produce biomass and microbial products, microbial cells may be used to catalyse the conversion of a compound into a structurally similar, but financially more valuable, compound. Such fermentations are termed transformation processes, biotransformations, or bioconversions. Although the production of vinegar is the oldest and most well-established transformation process (the conversion of ethanol into acetic acid), the majority of these processes involve the production of high-value compounds. Because microorganisms can behave as chiral catalysts with high regio- and stereo-specificity, microbial processes are more specific than purely chemical ones and make possible the addition, removal, or modification of functional groups at specific sites on a complex molecule without the use of chemical protection. The reactions that may be catalysed include oxidation, dehydrogenation, hydroxylation, dehydration and condensation, decarboxylation, deamination, animation, and isomerization. The anomaly of the transformation process is that a large biomass has to be produced to catalyse, perhaps, a single reaction. The logical development of these processes is to perform the reaction using the purified enzyme or the enzyme attached to an immobile support.

However, enzymes work more effectively within their microbial cells, especially if co-factors such as reduced pyridine nucleotide need to be regenerated. A compromise is to employ resting cells as catalysts, which may be suspended in a medium not supporting growth or attached to an immobile support. The reader is referred to Goodhue *et aZ.*** for a detailed review of transformation processes.

Recombinant Products

The advent of recombinant DNA technology has extended the range of potential microbial fermentation products. It is possible to introduce genes from higher organisms into microbial cells such that the recipient cells are capable of synthesizing foreign (or heterologous) proteins. Examples of the hosts for such foreign genes include *Escherichia coli, Saccharomyces cerevisiae* and other yeasts as well as filamentous fungi such as *Aspergillus niger* var. *awamori.*

Products produced in such genetically manipulated organisms include interferon, insulin, human serum albumin, factor VIII and factor IX, epidermal growth factor, bovine somatostatin and bovine chymosin. Important factors in the design of these processes include the secretion of the product, minimization of the degradation of the product, and the control of the onset of synthesis during the fermentation, as well as maximizing the expression of the foreign gene.

The Fermentation Process

Although the central component of the system is obviously the fermenter itself, in which the organism is grown under conditions optimum for product formation, one must not lose sight of operations upstream and downstream of the fermenter. Before the fermentation is started the medium must be formulated and sterilized, the fermenter sterilized, and a starter culture must be available in sufficient quantity and in the correct physiological state to inoculate the production fermenter. Downstream of the fermenter the product has to be purified and further processed and the effluents produced by the process have to be treated.

The Mode of Operation of Fermentation Processes

As discussed earlier, microorganisms may be grown in batch, fed-batch, or continuous culture, and continuous culture offers the most control over the growth of the cells. However, the commercial adoption of continuous culture is confined to the production of biomass and, to a limited extent, the production of potable and industrial alcohol. The superiority of continuous culture for biomass production is overwhelming, as may be seen from the following account, but for other microbial products the disadvantages of the system outweigh the improved process control which the technique offers.

The steady-state nature of a continuous process is also advantageous in that the system should be far easier to control than a comparable batch one. During a batch fermentation, heat output, acid or alkali production, and oxygen consumption will range from very low rates at the start of the fermentation to very high rates during the late logarithmic phase. Thus, the control of the environment of such a system is far more difficult than that of a continuous process where, at steady-state, production and consumption rates are constant. Furthermore, a continuous process should result in a more constant labour demand than a comparable batch one.

A frequently quoted disadvantage of continuous systems is their susceptibility to contamination by foreign organisms. The prevention of contamination is essentially a problem of fermenter design, construction, and operation and should be overcome by good engineering and microbiological practice. ICI recognized the overwhelming advantages of a continuous biomass process and overcame the problems of contam ination by building a secure fermenter capable of very long periods of aseptic operation, as described by Smith.

The production of growth-associated by-products, such as ethanol, should also be more efficient in continuous culture. However, continuous brewing has met with only limited success and UK breweries have abandoned such systems owing to problems of flavour and lack of production of industrial alcohol, on the other hand, should not be limited by the problems encountered by the brewing industry and continuous culture should be the method of choice for such a process. The adoption of continuous culture for the production of biosynthetic (as opposed to catabolic) microbial products has been extremely limited. Although, theoretically, it is possible to optimize a continuous system such that optimum productivity of a metabolite should be achieved, the long-term stability of such systems is precarious, owing to the problem of strain degeneration. A consideration of the kinetics of continuous culture reveals that the system is highly selective and will favour the propagation of the best-adapted organism in a culture. Best-adapted in this context refers to the affinity of the organism for the limiting substrate at the operating dilution rate. A commercial organism is usually highly mutated such that it will produce very high amounts of the desired product. Therefore, in physiological terms, such commercial organisms are extremely inefficient and a revertant strain, producing less of the desired product, may be better adapted to the cultural conditions than the superior producer and will come to dominate the culture. This phenomenon, termed by tamination from within, is the major reason for the lack of use of continuous culture for the production of microbial metabolites.

Although the fermentation industry has been reluctant to adopt continuous culture for the production of microbial metabolites, very considerable progress has been made in the development of fed-batch Fed-batch culture may be used to achieve a considerable degree of process control and to extend the productive period of a traditional batch process without the inherent disadvantages of continuous culture described previously. The major advantage of feeding a medium component to a culture, rather than incorporating it entirely in the initial batch, is that the nutrient may be maintained at a very low concentration during the fermentation. A low (but constantly replenished) nutrient level may be advantageous in:

(i) Maintaining conditions in the culture within the aeration capacity

(ii) Removing the repressive effects of medium components such as

(iii) Avoiding the toxic effects of a medium component.

(iv) Providing a limiting level of a required nutrient for an auxoof the fermenter. Rapidly used carbon and nitrogen sources and phosphate. Trophic strain.

The earliest example of the commercial use of fed-batch culture is the production of bakers' yeast. It was recognized as early as 1915 that an excess of malt in the production medium would result in a high rate of biomass production and an oxygen demand which could not be met by the fermenter. This resulted in the development of anaerobic conditions and the formation of ethanol at the expense of biomass. The solution to this problem was to grow the yeast initially in a weak medium and then add additional medium at a rate less than the organism could use it. It is now appreciated that a high glucose concentration represses respiratory activity, and in modern yeast production plants the feed of molasses is under strict control based on the automatic measurement of traces of ethanol in the exhaust gas of the fermenter. As soon as ethanol is detected the feed rate is reduced. Although such systems may result in low growth rates, the biomass yield is near that theoretically obtainable.

The penicillin fermentation provides a very good example of the use of fed-batch culture for the production of a secondary metabolite. The penicillin process is a two-stage fermentation; an initial growth phase is followed by the production phase or idiophase. During the production phase glucose is fed to the fermentation at a rate which allows a relatively high growth rate (and therefore rapid accumulation of biomass) yet maintains the oxygen demand of the culture within the aeration capacity of the equipment. If the oxygen demand of the biomass were to exceed the aeration capacity of the fermenter anaerobic conditions would result and the carbon source would be used inefficiently. During the production phase the biomass must be maintained at a relatively low growth rate and, thus, the glucose is fed at a low dilution rate. Phenylacetic acid is a precursor of the penicillin molecule but it is also toxic to the producer organism above a threshold concentration. Thus, the precursor is also fed into the fermentation continuously, thereby maintaining its concentration below the inhibitory level.

The Genetic Improvement of Product Formation

Owing to their inherent control systems, microorganisms usually produce commercially important metabolites in very low concentrations and, although the yield may be increased by optimizing the cultural conditions, productivity is controlled ultimately by the organism's genome. Thus, to improve the potential productivity, the organism's genome must be modified and this may be achieved in two ways: by (i) classical strain improvement by mutation and selection and (ii) the use of recombination.

Mutation

Each time a microbial cell divides there is a small probability of an inheritable change occurring. A strain exhibiting such a changed characteristic is termed a mutant and the process giving rise to it, a mutation.

The probability of a mutation occurring may be increased by exposing the culture to a mutagenic agent such as UV light, ionizing radiation, and various chemicals, for example nitrosoguanidine, nitrous acid and caffeine. Such an exposure usually involves subjecting the population to a mutagen dose which results in the death of the vast majority of the cells.

The survivors of the mutagen exposure may then contain some mutants, the vast majority of which will produce lower levels of the desired product. However, a very small proportion of the survivors may be improved producers. Thus, it is the task of the industrial geneticist to separate the desirable mutants (the superior producers) from the very many inferior types. This approach is easier for strains producing primary metabolites than it is for those producing secondary metabolites, as may be seen from the following examples.

The synthesis of a primary microbial metabolite (such as an amino acid) is controlled such that it is only produced at a level required by the organism. The control mechanisms involved are the inhibition of enzyme activity and the repression of enzyme synthesis by the end product when it is present in the cell at a sufficient concentration. Thus, these mechanisms are referred to as feedback control. It is obvious that a good 'commercial' mutant should lack the control systems so that 'overproduction' of the end product will result. The isolation of mutants of *Corynebacterium glutamicum* capable of producing lysine will be used to illustrate the approaches which have been adopted to remove the control systems.

A mutant which could not catalyse the conversion of aspartic semialdehyde into homoserine would be capable of growth only in a homoserine-supplemented medium and the organism would be described as a homoserine auxotroph. If such an organism were grown in the presence of very low concentrations of homoserine the endogenous level of threonine would not reach the inhibitory level for aspartokinase control and, thus, aspartate would be converted into lysine which would accumulate in the medium. Thus, a knowledge of the control of the biosynthetic pathway allows a 'blueprint' of the desirable mutant to be constructed and makes easier the task of designing the procedure to isolate the desired type from the other survivors of a mutation treatment.

The isolation of bacterial auxotrophs may be achieved using the penicillin enrichment technique developed by Davis. Under normal culture conditions an auxotroph is at a disadvantage compared with the parental (wild-type) cells. However, penicillin only kills growing cells and, therefore, if the survivors of a mutation treatment were cultured in a medium containing penicillin and lacking the growth requirement of the desired mutant only those cells unable to grow would survive, *i.e.* the desired auxotrophs. If the cells were removed from the penicillin broth, washed, and resuspended in a medium containing the requirement of the desired auxotroph then the resulting culture should be rich in the required type. Nakayama this technique to isolate a homoserine auxotroph of C. *glutamicum* which produced 44 g 1-' lysine.

An alternative approach to the isolation of mutants which do not produce controlling end products *(i.e.* auxotrophs) is to isolate mutants which do not recognize the presence of controlling compounds. Such mutants may be isolated from the survivors of a mutation treatment by exploiting their capacity to grow in the presence of certain compounds which are inhibitory to the parental types. An analogue is a compound which is similar in structure to another compound and analogues of primary metabolites are frequently inhibitory to microbial cells. The toxicity of the analogue may be due to any of a number of possible mechanisms; for example, the analogue may be incorporated into a macromolecule in place of the natural product, resulting in the production of a defective compound, or the analogue may act as a competitive inhibitor of an enzyme for which the natural product is a substrate. Also, the analogue may mimic the control characteristics of the natural product and inhibit product formation despite the fact that the natural product concentration is inadequate to support growth. A mutant which is capable of growing in the presence of an analogue inhibitory to the parent may owe its resistance to any of a number of mechanisms.

However, if the toxicity were due to the analogue mimicking the control characteristics of the normal end product, then the resistance may be due to the control system being unable to recognize the analogue as a control factor. Such analogue-resistant mutants may also not recognize the natural product and may, therefore, overproduce it. Thus, there is a reasonable probability that mutants resistant to the inhibitory effects of an analogue may overproduce the compound to which the analogue is analogous. Sano and approach in attempting to isolate lysine-producing mutants of *Brevibacterium Jiavum.*

Thus, a knowledge of the control systems may assist in the design of procedures for the isolation of mutants overproducing primary metabolites.

The design of procedures for the isolation of mutants overproducing secondary metabolites has been more difficult owing to the fact that far less information was available on the control of production and, also, that the end products of secondary metabolism are not required for growth. Thus, many current industrial strains have been selected using direct, empirical, screens of the survivors of a mutation treatment for productivity rather than cultural systems which give an advantage to potential superior producers.

Microbial Enzymes

Microbial enzymes are most widely used in the food and beverage industries and to a lesser extent in clinical and analytical laboratories and as protease detergents in washing powders. The most economical and convenient method of producing these enzymes is by microbial fermentation. Bacillus stearothermophilusproduces amylases as secondary metabolites, but most other microbes produce enzymes as primary metabolites, during exponential growth.

The major commercial utilization of microbial enzymes is in the food and beverage industries (Jeenes et al.,1991), in pharmaceutical textile and leather industries, in analytical processes as well as in detergent and dairy industry. Most enzymes are synthesized in the logarithmic phase of the batch culture and may, therefore, be considered as primary metabolites.

However, some, for example, amylases of Bacillus stearothemophilus are produced by idiophase, type cultures and may be considered as equivalent to secondary metabolites (Manning and Campbell, 1961). Enzymes may be produced from animals and plants as well as microbial sources, but the production by microbial fermentation is the most economical and convenient method. For example, microbial rennin is being used instead of calf rennin and proteases used in detergents are produced by microbes. It is now possible to engineer microbial cells to produce animal or plant enzymes.

Most of the enzymes in industrial use are extracellular proteins produced by Aspergillus sp. or Bacillussp. and include α-amylase, β-glucanase, cellulase, dextranase, lactase, lipase, pectinase, proteases and others. The extent of purification required for most of these enzymes is minimal, and they can be produced in tonnes without serious problems. Some of the important microbial enzymes, their source and applications

Transformation Process

The microbial cells can be used to catalyse the conversion of a compound into a structurally similar, but financially more valuable compound. Such fermentations are termed transformation processes, biotransformation, or bioconversions. The production of vinegar is the oldest and most well established transformation process (the conversion of ethanol into acetic acid). Other examples of transformation are sorbitol to sorbose and production of new antibiotics that are more effective than the existing ones. The anomaly of the transformation process is that a large biomass has to be produced to catalyse a single reaction.

Recombinant Products

The advent of recombinant DNA technology has extended the range of potential microbial fermentation products. It is possible to introduce genes from higher organisms into microbial cells such that the recipient cells are capable of synthesizing foreign (or heterologous) proteins.

Examples of the 'hosts' for such foreign genes include E. coli, Saccharomyces cerevisiaeand other yeasts and fungi. Products produced by genetically manipulated organisms include interferon, insulin, human serum albumin, factors VIII and IX, epidermal growth factor and bovine somatostatin.

Brewing Process of Beer

Beer is the world's most widely consumed and probably the oldest of alcoholic beverages; it is the third most popular drink overall, after water and tea. It is produced by the brewing and fermentation of starches, mainly derived from cereal grains—most commonly malted barley, although wheat, maize (corn), and rice are widely used. Most beer is flavoured with hops, which add bitterness and act as a natural preservative, though other flavourings such as herbs or fruit may occasionally be included. Some of humanity's earliest known writings refer to the production and distribution of beer: the Code of Hammurabi included laws regulating beer and beer parlours, and "The Hymn to Ninkasi", a prayer to the Mesopotamian goddess of beer, served as both a prayer and as a method of remembering the recipe for beer in a culture with few literate people. Today, the brewing industry is a global business, consisting of several dominant multinational companies and many thousands of smaller producers ranging from brewpubs to regional breweries.

The strength of beer is usually around 4% to 6% alcohol by volume (abv) though may range from less than 1% abv, to over 20% abv in rare cases.

Beer forms part of the culture of beer-drinking nations and is associated with social traditions such as beer festivals, as well as a rich pub culture involving activities like pub crawling and pub games such as bar billiards.

History

Beer is one of the world's oldest prepared beverages, possibly dating back to the early Neolithic or 9500 BC, when cereal was first farmed, and is recorded in the written history of ancient Egypt and Mesopotamia. Archaeologists speculate that beer was instrumental in the formation of civilisations.

The earliest known chemical evidence of beer dates to circa 3500–3100 BC from the site of Godin Tepe in the Zagros Mountains of western Iran. Some of the earliest Sumerian writings found in the region contain references to a type of beer; one such example, a prayer to the goddess Ninkasi, known as "The Hymn to Ninkasi", served as both a prayer as well as a method of remembering the recipe for beer in a culture with few literate people. The Ebla tablets, discovered in 1974 in Ebla, Syria and date back to 2500 BC, reveal that the city produced a range of beers, including one that appears to be named "Ebla" after the city. A beer made from rice, which, unlike sake, didn't use the amylolytic process, and was probably prepared for fermentation by mastication or malting, was made in China around 7000 BC.

As almost any substance containing carbohydrates, mainly sugars or starch, can naturally undergo fermentation, it is likely that beer-like beverages were independently invented among various cultures throughout the world. The invention of bread and beer has been argued to be responsible for humanity's ability to develop technology and build civilization.

Beer was spread through Europe by Germanic and Celtic tribes as far back as 3000 BC, and it was mainly brewed on a domestic scale. The product that the early Europeans drank might not be recognised as beer by most people today. Alongside the basic starch source, the early European beers might contain fruits, honey, numerous types of plants, spices and other substances such as narcotic herbs. What they did not contain was hops, as that was a later addition first mentioned in Europe around 822 by a Carolingian Abbot and again in 1067 by Abbess Hildegard of Bingen.

In 1516, William IV, Duke of Bavaria, adopted the *Reinheitsgebot* (purity law), perhaps the oldest food-quality regulation still in use through the 20th century, according to which the only allowed ingredients of beer are water, hops and barley-malt. Beer produced before the Industrial Revolution continued to be made and sold on a domestic scale, although by the 7th century AD, beer was also being produced and sold by European monasteries. During the Industrial Revolution, the production of beer moved from artisanal manufacture to industrial manufacture, and domestic manufacture ceased to be significant by the end of the 19th century. The development of hydrometers and thermometers changed brewing by allowing the brewer more control of the process and greater knowledge of the results.

Today, the brewing industry is a global business, consisting of several dominant multinational companies and many thousands of smaller producers ranging from brewpubs to regional breweries. As of 2006, more than 133 billion litres (35 billion gallons), the equivalent of a cube 510 metres on a side, of beer are sold per year, producing total global revenues of $294.5 billion (£147.7 billion).

Brewing

The process of making beer is known as brewing. A dedicated building for the making of beer is called a brewery, though beer can be made in the home and has been for much of its history. A company that makes beer is called either a brewery or a brewing company. Beer made on a domestic scale for non-commercial reasons is classified as homebrewing regardless of where it is made, though most homebrewed beer is made in the home. Brewing beer is subject to legislation and taxation in developed countries, which from the late 19th century largely restricted brewing to a commercial operation only. However, the UK government relaxed legislation in 1963, followed by Australia in 1972 and the USA in 1979, allowing homebrewing to become a popular hobby.

The purpose of brewing is to convert the starch source into a sugary liquid called wort and to convert the wort into the alcoholic beverage known as beer in a fermentation process effected by yeast.

The sweet wort collected from sparging is put into a kettle, or "copper", (so called because these vessels were traditionally made from copper) and boiled, usually for about one hour. During boiling, water in the wort evaporates, but the sugars and other components of the wort remain; this allows more efficient use of the starch sources in the beer. Boiling also destroys any remaining enzymes left over from the mashing stage. Hops are added during boiling as a source of bitterness, flavour and aroma. Hops may be added at more than one point during the boil. The longer the hops are boiled, the more bitterness they contribute, but the less hop flavour and aroma remains in the beer.

After boiling, the hopped wort is now cooled, ready for the yeast. In some breweries, the hopped wort may pass through a hopback, which is a small vat filled with hops, to add aromatic hop flavouring and to act as a filter; but usually the hopped wort is simply cooled for the fermenter, where the yeast is added. During fermentation, the wort becomes beer in a process which requires a week to months depending on the type of yeast and strength of the beer. In addition to producing alcohol, fine particulate matter suspended in the wort settles during fermentation. Once fermentation is complete, the yeast also settles, leaving the beer clear.

Fermentation is sometimes carried out in two stages, primary and secondary. Once most of the alcohol has been produced during primary fermentation, the beer is transferred to a new vessel and allowed a period of secondary fermentation. Secondary fermentation is used when the beer requires long storage before packaging or greater clarity. When the beer has fermented, it is packaged either into casks for cask ale or kegs, aluminium cans, or bottles for other sorts of beer.

Ingredients

The basic ingredients of beer are water; a starch source, such as malted barley, able to be fermented (converted into alcohol); a brewer's yeast to produce the fermentation; and a flavouring such as hops. A mixture of starch sources may be used, with a secondary starch source, such as maize (corn), rice or sugar, often being termed an adjunct, especially when used as a lower-cost substitute for malted barley. Less widely used starch sources include millet, sorghum and cassava root in Africa, potato in Brazil, and agave in Mexico, among others. The amount of each starch source in a beer recipe is collectively called the grain bill.

Water

Beer is composed mostly of water. Regions have water with different mineral components; as a result, different regions were originally better suited to making certain types of beer, thus giving them a regional character. For example, Dublin has hard water well suited to making stout, such as Guinness; while Pilzen has soft water well suited to making pale lager, such as Pilsner Urquell. The waters of Burton in England contain gypsum, which benefits making pale ale to such a degree that brewers of pale ales will add gypsum to the local water in a process known as Burtonisation.

Starch Source

The starch source in a beer provides the fermentable material and is a key determinant of the strength and flavour of the beer. The most common starch source used in beer is malted grain. Grain is malted by soaking it in water, allowing it to begin germination, and then drying the partially germinated grain in a kiln. Malting grain produces enzymes that convert starches in the grain into fermentable sugars. Different roasting times and temperatures are used to produce different colours of malt from the same grain. Darker malts will produce darker beers.

Nearly all beer includes barley malt as the majority of the starch. This is because of its fibrous husk, which is not only important in the sparging stage of brewing (in which water is washed over the mashed barley grains to form the wort), but also as a rich source of amylase, a digestive enzyme which facilitates conversion of starch into sugars. Other malted and unmalted grains (including wheat, rice, oats, and rye, and less frequently, corn and sorghum) may be used. In recent years, a few brewers have produced gluten-free beer made with sorghum with no barley malt for those who cannot consume gluten-containing grains like wheat, barley, and rye.

Hops

Flavouring beer is the sole major commercial use of hops. The flower of the hop vine is used as a flavouring and preservative agent in nearly all beer made today. The flowers themselves are often called "hops".

Hops were used by monastery breweries, such as Corvey in Westphalia, Germany, from AD 822, though the date normally given for widespread cultivation of hops for use in beer is the thirteenth century. Before the thirteenth century, and until the sixteenth century, during which hops took over as the dominant flavouring, beer was flavoured with other plants; for instance, *Glechoma hederacea*. Combinations of various aromatic herbs, berries, and even ingredients like wormwood would be combined into a mixture known as gruit and used as hops are now used. Some beers today, such as Fraoch' by the Scottish Heather Ales company and Cervoise Lancelot by the French Brasserie-Lancelot company, use plants other than hops for flavouring.

Hops contain several characteristics that brewers desire in beer. Hops contribute a bitterness that balances the sweetness of the malt; the bitterness of beers is measured on the International Bitterness Units scale. Hops contribute floral, citrus, and herbal aromas and flavours to beer. Hops have an antibiotic effect that favours the activity of brewer's yeast over less desirable microorganisms, and hops aids in "head retention", the length of time that a foamy head created by carbonation will last. The acidity of hops is a preservative.

Yeast

Yeast is the microorganism that is responsible for fermentation in beer. Yeast metabolises the sugars extracted from grains, which produces alcohol and carbon dioxide, and thereby turns wort into beer. In addition to fermenting the beer, yeast influences the character and flavour. The dominant types of yeast used to make beer are the top-fermenting *Saccharomyces cerevisiae* and bottom-fermenting *Saccharomyces uvarum*. *Brettanomyces* ferments lambics, and *Torulaspora delbrueckii* ferments Bavarian weissbier. Before the role of yeast in fermentation was understood, fermentation involved wild or airborne yeasts. A few styles such as lambics rely on this method today, but most modern fermentation adds pure yeast cultures.

Clarifying Agent

Some brewers add one or more clarifying agents to beer, which typically precipitate (collect as a solid) out of the beer along with protein solids and are found only in trace amounts in the finished product. This process makes the beer appear bright and clean, rather than the cloudy appearance of ethnic and older styles of beer such as wheat beers. Examples of clarifying agents include isinglass, obtained from swimbladders of fish; Irish moss, a seaweed; kappa carrageenan, from the seaweed *Kappaphycus cottonii*; Polyclar (artificial); and gelatin. If a beer is marked "suitable for Vegans", it was clarified either with seaweed or with artificial agents.

Production

The brewing industry is a global business, consisting of several dominant multinational companies and many thousands of smaller producers ranging from brewpubs to regional breweries. More than 133 billion litres (35 billion gallons) are sold per year—producing total global revenues of $294.5 billion (£147.7 billion) in 2006.

A microbrewery, or craft brewery, is a modern brewery which produces a limited amount of beer. The maximum amount of beer a brewery can produce and still be classed as a microbrewery varies by region and by authority, though is usually around 15,000 barrels (18,000 hectolitres/ 475,000 US gallons) a year. A brewpub is a type of microbrewery that incorporates a pub or other eating establishment. SABMiller became the largest brewing company in the world when it acquired Royal Grolsch, brewer of Dutch premium beer brand Grolsch. InBev was the second-largest beer-producing company in the world, and Anheuser-Busch held the third spot, but after the merger between InBev and Anheuser-Busch, the new Anheuser-Busch InBev company is the largest brewer in the world. Brewing at home is subject to regulation and prohibition in many countries. Restrictions on homebrewing were lifted in the UK in 1963, Australia followed suit in 1972, and the USA in 1978, though individual states were allowed to pass their own laws limiting production.

Varieties

While there are many types of beer brewed, the basics of brewing beer are shared across national and cultural boundaries. The traditional European brewing regions—Germany, Belgium, the United Kingdom, Ireland, Poland, the Czech Republic, Scandinavia, the Netherlands and Austria—have local varieties of beer. In some countries, notably the USA, Canada, and Australia, brewers have adapted European styles to such an extent that they have effectively created their own indigenous types.

Michael Jackson, in his 1977 book *The World Guide To Beer*, categorised beers from around the world in local style groups suggested by local customs and names. Fred Eckhardt furthered Jackson's work in *The Essentials of Beer Style* in 1989.

Top-fermented beers are most commonly produced with *Saccharomyces cerevisiae* which clumps and rises to the surface, typically between 15 and 24°C (60 and 75°F). At these temperatures, yeast produces significant amounts of esters and other secondary flavour and aroma products, and the result is often a beer with slightly "fruity" compounds resembling apple, pear, pineapple, banana, plum, or prune, among others. Before the introduction of hops into England from the Netherlands in the 15th century, the name "ale" was exclusively applied to unhopped fermented beverages, the term *beer* being gradually introduced to describe a brew with an infusion of hops. The word *ale* may come from the Old English *ealu*, in turn from the Proto-Indo-European base **alut-*, which holds connotations of "sorcery, magic, possession, intoxication". Real ale is the term coined by the Campaign for Real Ale (CAMRA) in 1973 for "beer brewed from traditional ingredients, matured by secondary fermentation in the container from which it is dispensed, and served without the use of extraneous carbon dioxide". It is applied to bottle conditioned and cask conditioned beers.

Pale Ale

Pale ale is a beer which uses a top-fermenting yeast and predominantly pale malt. It is one of the world's major beer styles.

Stout

Stout and porter are dark beers made using roasted malts or roast barley, and typically brewed with slow fermenting yeast. There are a number of variations including Baltic porter, dry stout, and Imperial stout. The name Porter was first used in 1721 to describe a dark brown beer popular with the street and river porters of London. This same beer later also became known as stout, though the word stout had been used as early as 1677. The history and development of stout and porter are intertwined.

Mild

Mild ale has a predominantly malty palate. It is usually dark coloured with an abv of 3% to 3.6%, although there are lighter hued milds as well as stronger examples reaching 6% abv and higher.

Wheat

Wheat beer is brewed with a large proportion of wheat although it often also contains a significant proportion of malted barley. Wheat beers are usually top-fermented (in Germany they have to be by law). The flavour of wheat beers varies considerably, depending upon the specific style.

Lager

Lager is the English name for cool fermenting beers of Central European origin. Pale lagers are the most commonly consumed beers in the world. The name "lager" comes from the German "lagern" for "to store", as brewers around Bavaria stored beer in cool cellars and caves during the warm summer months. These brewers noticed that the beers continued to ferment, and to also clear of sediment, when stored in cool conditions. Lager yeast is a cool bottom-fermenting yeast (*Saccharomyces pastorianus*) and typically undergoes primary fermentation at 7–12 °C (45–54 °F) (the fermentation phase), and then is given a long secondary fermentation at 0–4 °C (32–39 °F) (the lagering phase). During the secondary stage, the lager clears and mellows. The cooler conditions also inhibit the natural production of esters and other byproducts, resulting in a "cleaner"-tasting beer.

Modern methods of producing lager were pioneered by Gabriel Sedlmayr the Younger, who perfected dark brown lagers at the Spaten Brewery in Bavaria, and Anton Dreher, who began brewing a lager (now known as Vienna lager), probably of amber-red colour, in Vienna in 1840–1841. With improved modern yeast strains, most lager breweries use only short periods of cold storage, typically 1–3 weeks.

Lambic

Lambic, a beer of Belgium, is naturally fermented using wild yeasts, rather than cultivated. Many of these are not strains of brewer's yeast (*Saccharomyces cerevisiae*) and may have significant differences in aroma and sourness. Yeast varieties such as *Brettanomyces bruxellensis* and *Brettanomyces lambicus* are common in lambics. In addition, other organisms such as Lactobacillus bacteria produce acids which contribute to the sourness.

Measurement

Beer is measured and assessed by bitterness, by strength and by colour. The perceived bitterness is measured by the International Bitterness Units scale (IBU), defined in co-operation between the American Society of Brewing Chemists and the European Brewery Convention. The international scale was a development of the European Bitterness Units scale, often abbreviated as EBU, and the bitterness values should be identical.

Colour

Beer colour is determined by the malt. The most common colour is a pale amber produced from using pale malts. *Pale lager* and *pale ale* are terms used for beers made from malt dried with coke. Coke was first used for roasting malt in 1642, but it was not until around 1703 that the term *pale ale* was used. In terms of sales volume, most of today's beer is based on the pale lager brewed in 1842 in the town of Pilsen in the present-day Czech Republic. The modern pale lager is light in colour with a noticeable carbonation (fizzy bubbles) and a typical alcohol by volume content of around 5%. The Pilsner Urquell, Bitburger, and Heineken brands of beer are typical examples of pale lager, as are the American brands Budweiser, Coors, and Miller. Dark beers are usually brewed from a pale malt or lager malt base with a small proportion of darker malt added to achieve the desired shade. Other colourants—such as caramel—are also widely used to darken beers. Very dark beers, such as stout, use dark or patent malts that have been roasted longer. Some have roasted unmalted barley.

Strength

Beer ranges from less than 3% alcohol by volume (abv) to around 14% abv, though this strength can be increased to around 20% by re-pitching with champagne yeast, and to 55% abv by the freeze-distilling process. The alcohol content of beer varies by local practice or beer style. The pale lagers that most consumers are familiar with fall in the range of 4–6%, with a typical abv of 5%. The customary strength of British ales is quite low, with many session beers being around 4% abv. Some beers, such as table beer are of such low alcohol content (1%–4%) that they are served instead of soft drinks in some schools.

The alcohol in beer comes primarily from the metabolism of sugars that are produced during fermentation. The quantity of fermentable sugars in the wort and the variety of yeast used to ferment the wort are the primary factors that determine the amount of alcohol in the final beer.

Additional fermentable sugars are sometimes added to increase alcohol content, and enzymes are often added to the wort for certain styles of beer (primarily "light" beers) to convert more complex carbohydrates (starches) to fermentable sugars. Alcohol is a by-product of yeast metabolism and is toxic to the yeast; typical brewing yeast cannot survive at alcohol concentrations above 12% by volume. Low temperatures and too little fermentation time decreases the effectiveness of yeasts and consequently decreases the alcohol content.

Exceptionally Strong Beers

The strength of beers has climbed during the later years of the 20th century. Vetter 33, a 10.5% abv (33 degrees Plato, hence Vetter "33") doppelbock, was listed in the 1994 *Guinness Book of World Records* as the strongest beer at that time, though Samichlaus, by the Swiss brewer Hurlimann, had also been listed by the *Guinness Book of World Records* as the strongest at 14% abv. Since then, some brewers have used champagne yeasts to increase the alcohol content of their beers. Samuel Adams reached 20% abv with *Millennium*, and then surpassed that amount to 25.6% abv with Utopias. The strongest beer brewed in Britain was Baz's Super Brew by Parish Brewery, a 23% abv beer. The product that is claimed to be the strongest beer made is *The End of History*, a 55% Belgian ale, made by the Scottish brewery BrewDog in 2010, who also made *Sink The Bismarck!*, a 41% abv IPA, and *Tactical Nuclear Penguin*, a 32% abv Imperial Stout; these are made using the eisbock method of freeze distilling in which the brewery freeze distils a strong ale, gradually removing the ice and freezing again until the beer reaches the strength required, a process that may class the finished products as spirits rather than beer. The German brewery Schorschbräu's *Schorschbock*—a 31% abv eisbock, and Hair of the Dog's *Dave*—a 29% abv barley wine made in 1994, both used the same freeze distilling method. A 60% abv blend of beer with whiskey was jokily claimed as the strongest beer by a Dutch brewery in July 2010.

Packaging

Most beers are cleared of yeast by filtering when packaged in bottles and cans. However, bottle conditioned beers retain some yeast—either by being unfiltered, or by being filtered and then reseeded with fresh yeast. It is usually recommended that the beer be poured slowly, leaving any yeast sediment at the bottom of the bottle. However, some drinkers prefer to pour in the yeast; this practice is customary with wheat beers. Typically, when serving a hefeweizen, 90% of the contents are poured, and the remainder is swirled to suspend the sediment before pouring it into the glass. Alternatively, the bottle may be inverted prior to opening. Glass bottles are always used for bottle conditioned beers. Many beers are sold in cans, though there is considerable variation in the proportion between different countries. In Sweden in 2001, 63.9% of beer was sold in cans. People either drink from the can or pour the beer into a glass. Cans protect the beer from light (thereby preventing "skunked" beer) and have a seal less prone to leaking over time than bottles. Cans were initially viewed as a technological breakthrough for maintaining the quality of a beer, then became commonly associated with less expensive, mass-produced beers, even though the quality of storage in cans is much like bottles. Plastic (PET) bottles are used by some breweries.

Temperature

The temperature of a beer has an influence on a drinker's experience; warmer temperatures reveal the range of flavours in a beer but cooler temperatures are more refreshing. Most drinkers prefer pale lager to be served chilled, a low- or medium-strength pale ale to be served cool, while a strong barley wine or imperial stout to be served at room temperature.Beer writer Michael Jackson proposed a five-level scale for serving temperatures: well chilled (7 °C/45 °F) for "light" beers (pale lagers); chilled (8 °C/46 °F) for Berliner Weisse and other wheat beers; lightly chilled (9 °C/48 °F) for all dark lagers, altbier and German wheat beers; cellar temperature (13 °C/55 °F) for regular British ale, stout and most Belgian specialities; and room temperature (15.5 °C/59.9 °F) for strong dark ales (especially trappist beer) and barley wine. Drinking chilled beer began with the development of artificial refrigeration and by the 1990s, was spread in those countries that concentrated on brewing pale lager. Chilling beer makes it more refreshing, though below 15.5 °C (59.9 °F) the chilling starts to reduce taste awareness and reduces it significantly below 10 °C (50 °F). Beer served unchilled—either cool or at room temperature, reveal more of their flavours. Cask Marque, a non-profit UK beer organisation, has set a temperature standard range of 12°-14 °C (53°-57 °F) for cask ales to be served.

Vessels

Beer is consumed out of a variety of vessels, such as a glass, a beer stein, a mug, a pewter tankard, a beer bottle or a can. The shape of the glass from which beer is consumed can influence the perception of the beer and can define and accent the character of the style. Breweries offer branded glassware intended only for their own beers as a marketing promotion, as this increases sales. The pouring process has an influence on a beer's presentation. The rate of flow from the tap or other serving vessel, tilt of the glass, and position of the pour (in the centre or down the side) into the glass all influence the end result, such as the size and longevity of the head, lacing (the pattern left by the head as it moves down the glass as the beer is drunk), and turbulence of the beer and its release of carbonation.

2

Fermentation Process of Grape Juice for Wine Production

Wine is Fermented Grape Juice

Wine can be made from grapes, fruits, berries etc. Most wine, though, is made from grapes. And no matter what the wine is made from, there must be fermentation, that is, that sugar be transformed into alcohol. If the amount of alcohol is relatively low, the result is wine. If it is high, the result is a "distilled liquor," something like gin or vodka

There are red wines, pink wines (also known as "rose" or some-times "blush") and white wines.

Red wine result when the crushed grape skin pulp and seeds of purple or red varieties are allowed to remain with juice during fermentation periods.

Pink/rose wine can be produced by removing the non-juice pumace from the must during fermentation.

White wines can be made from pigmented grapes by removal of skins, pulp and seeds before juice fermentation.

Wines might be "fortified," "sparkling," or "table."

In fortified wines, brandy is added to make the alcohol content higher (around 14 to 30 percent). These are less perishable and may be stable without pasteurization.

Wines are termed still or sparkling depending upon the amount of CO_2 they contain. The carbon dioxide may be formed naturally during fermentation or may be added artificially.

Both table and sparkling wines tend to have alcohol contents between 7 and 14 percent.

Sparkling wines are the ones with bubbles (greater CO_2), like Champagne.

Table wine (which can also be called "still") are the most "natural". The alcohol concentration itself is not sufficient to preserve natural wines, they are pasteurized.

The term light wine is also used to describe wine having alcohol content from 5- 10 %.

How is Wine Made?

Growing Grapes

Grapes grow on vines. There are many different types of grapes, but the best wine grape is the European Vitis vinifera. It is considered optimal because it has the right balance of sugar and acid to create a good fermented wine without the addition of sugar or water.

Harvest

Weather is a major factor is determining whether a year is going to be a "good vintage" (or "year"). For example, was there enough heat during the growing season to lead to enough sugar? At harvest time, the short-term effects of weather are quite important. To produce great wine, the fruit should have a high (but not overly high) sugar content ("brix"). Think of raisins.

As the fruit dries, the water evaporates. What is left is the sugary fruit. If it rains just at the point the wine grapes are ready, and before the grapes can be harvested, the additional water will cause the water level to increase, and the brix will go down. Not good. (You might ask, why not just add some sugar in the wine making process? Some do. Also considered "not good.")

Every year the wine grape grower plays a game of chance and must decide when to harvest. Simplistically, if you knew it wasn't going to rain, you would just test the brix until it was just right, then harvest. If you harvest too soon, you will probably end up getting a wine too low in alcohol content (there won't have been enough sugar to convert to alcohol). These wines will be "thin." If you delay harvest, there may be too much sugar, which leads to too low acid content. This also affects the taste (and the aging possibilities) of the wine.

Initial Processing of the Grape Juice

Grapes can (and might still) be crushed by stomping on them with your feet in a big vat. But a more practical way is to use a machine which does the job (and at the same time, removes the stems).

What you get may or may not get immediately separated. Skin and seeds might immediately be removed from the juice. Separation may not immediately occur (especially for red wines), since skins and stems are an important source of "tannins" which affect wine's taste and maturity through aging. The skins also determine the colour of the wine.

Maceration (the time spent while skins and seeds are left with the juice) will go on for a few hours or a few weeks. Pressing will then occur. One way to press the grapes is to use a "bladder press," a large cylindrical container that contains bags that are inflated and deflated several times, each time gently squeezing the grapes until all the juice has run free, leaving behind the rest of the grapes. You can also separate solids from juice through the use of a centrifuge.

Crushing and Primary Fermentation

Crushing is the process of gently squeezing the berries and breaking the skins to start to liberate the contents of the berries. Desteming is the process of removing the grapes from the rachis (the stem which holds the grapes). In traditional and smaller-scale wine making, the harvested grapes are sometimes crushed by trampling them barefoot or by the use of inexpensive small scale crushers. These can also destem at the same time. However, in larger wineries, a mechanical crusher/destemmer is used. The decision about desteming is different for red and white wine making. Generally when making white wine the fruit is only crushed, the stems are then placed in the press with the berries. The presence of stems in the mix facilitates pressing by allowing juice to flow past flattened skins. These accumulate at the edge of the press. For red winemaking, stems of the grapes are usually removed before fermentation since the stems have a relatively high tannin content; in addition to tannin they can also give the wine a vegetal aroma (due to extraction of 2-methoxy-3-isopropylpyrazine which has an aroma reminiscent of green bell peppers.) On occasion, the winemaker may decide to leave them in if the grapes themselves contain less tannin than desired. This is more acceptable if the stems have 'ripened' and started to turn brown. If increased skin extraction is desired, a winemaker might choose to crush the grapes after destemming. Removal of stems first means no stem tannin can be extracted. In these cases the grapes pass between two rollers which squeeze the grapes enough to separate the skin and pulp, but not so much as to cause excessive shearing or tearing of the skin tissues. In some cases, notably with "delicate" red varietals such as Pinot noir or Syrah, all or part of the grapes might be left uncrushed (called "whole berry") to encourage the retention of fruity aromas through partial carbonic maceration.

Most red wines derive their colour from grape skins (the exception being varieties or hybrids of non-vinifera vines which contain juice pigmented with the dark Malvidin 3,5-diglucoside anthocyanin) and therefore contact between the juice and skins is essential for colour extraction. Red wines are produced by destemming and crushing the grapes into a tank and leaving the skins in contact with the juice throughout the fermentation (maceration). It is possible to produce white (colourless) wines from red grapes by the fastidious pressing of uncrushed fruit. This minimizes contact between grape juice and skins (as in the making of *Blanc de noirs* sparkling wine, which is derived from Pinot noir, a red vinifera grape.)

Most white wines are processed without destemming or crushing and are transferred from picking bins directly to the press. This is to avoid any extraction of tannin from either the skins or grapeseeds, as well as maintaining proper juice flow through a matrix of grape clusters rather than loose berries. In some circumstances winemakers choose to crush white grapes for a short period of skin contact, usually for three to 24 hours. This serves to extract flavour and tannin from the skins (the tannin being extracted to encourage protein precipitation without excessive Bentonite addition) as well as Potassium ions, which participate in bitartrate precipitation (cream of tartar). It also results in an increase in the pH of the juice which may be desirable for overly acidic grapes. This was a practice more common in the 1970s than today, though still practiced by some Sauvignon blanc and Chardonnay producers in California.

In the case of rose wines, the fruit is crushed and the dark skins are left in contact with the juice just long enough to extract the colour that the winemaker desires. The must is then pressed, and fermentation continues as if the wine maker was making a white wine.

Yeast is normally already present on the grapes, often visible as a powdery appearance of the grapes. The fermentation can be done with this natural yeast, but since this can give unpredictable results depending on the exact types of yeast that are present, cultured yeast is often added to the must. One of the main problems with the use of wild ferments is the failure for the fermentation to go to completion, that is some sugar remains unfermented. This can make the wine sweet when a dry wine is desired. Frequently wild ferments lead to the production of unpleasant acetic acid (vinegar) production as a by product.

During the primary fermentation, the yeast cells feed on the sugars in the must and multiply, producing carbon dioxide gas and alcohol. The temperature during the fermentation affects both the taste of the end product, as well as the speed of the fermentation. For red wines, the temperature is typically 22 to 25 °C, and for white wines 15 to 18 °C. For every gram of sugar that is converted, about half a gram of alcohol is produced, so to achieve a 12% alcohol concentration, the must should contain about 24% sugars. The sugar percentage of the must is calculated from the measured density, the must weight, with the help of a specialized type of hydrometer called a saccharometer. If the sugar content of the grapes is too low to obtain the desired alcohol percentage, sugar can be added (chaptalization). In commercial winemaking, chaptalization is subject to local regulations.

During or after the alcoholic fermentation, malolactic fermentation can also take place, during which specific strains of bacteria convert malic acid into the milder lactic acid. This fermentation is often initiated by inoculation with desired bacteria.

Pressing

Pressing is the act of applying pressure to grapes or pomace in order to separate juice or wine from grapes and grape skins. Pressing is not always a necessary act in winemaking; if grapes are crushed there is a considerable amount of juice immediately liberated (called free-run juice) that can be used for vinification. Typically this free-run juice is of a higher quality than the press juice. However, most wineries do use presses in order to increase their production (gallons) per ton, as pressed juice can represent between 15%-30% of the total juice volume from the grape.

Presses act by positioning the grape skins or whole grape clusters between a rigid surface and a moveable surface and slowly decrease the volume between the two surfaces. Modern presses dictate the duration and pressure at each press cycle, usually ramping from 0 Bar to 2.0 Bar. Sometimes winemakers choose pressures which separate the streams of pressed juice, called making "press cuts." As the pressure increases the amount of tannin extracted from the skins into the juice increases, often rendering the pressed juice excessively tannic or harsh. Because of the location of grape juice constituents in the berry (water and acid are found primarily in the mesocarp or pulp, whereas tannins are found primarily in the pericarp, or skin, and seeds), pressed juice or wine tends to be lower in acidity with a higher pH than the free-run juice.

Before the advent of modern winemaking, most presses were basket presses made of wood and operated manually. Basket presses are composed of a cylinder of wooden slats on top of a fixed plate, with a moveable plate that can be forced downward (usually by a central ratcheting threaded screw.) The press operator would load the grapes or pomace into the wooden cylinder, put the top plate in place and lower it until juice flowed from the wooden slats. As the juice flow decreased, the plate was ratcheted down again. This process continued until the press operator determined that the quality of the pressed juice or wine was below standard, or all liquids had been pressed. Since the early 1990s, modern mechanical basket presses have been revived through higher-end producers seeking to replicate the gentle pressing of the historical basket presses. Because basket presses have a relatively compact design, the press cake offers a relatively longer pathway for the juice to travel before leaving the press. It is believed by advocates of basket presses that this relatively long pathway through the grape or pomace cake serves as a filter to solids that would otherwise affect the quality of the press juice. With red wines, the must is pressed after primary fermentation, which separates the skins and other solid matter from the liquid. With white wine, the liquid is separated from the must before fermentation. With rose, the skins may be kept in contact for a shorter period to give colour to the wine, in that case the must may be pressed as well. After a period in which the wine stands or ages, the wine is separated from the dead yeast and any solids that remained (called lees), and transferred to a new container where any additional fermentation may take place.

Pigeage

Pigeage is a French winemaking term for the traditional stomping of grapes in open fermentation tanks. To make certain types of wine, grapes are put through a crusher and then poured into open fermentation tanks. Once fermentation begins, the grape skins are pushed to the surface by carbon dioxide gases released in the fermentation process. This layer of skins and other solids is known as the cap. As the skins are the source of the tannins, the cap needs to be mixed through the liquid each day, or "punched," which traditionally is done by stomping through the vat.

Cold and Heat Stabilization

Cold stabilization is a process used in winemaking to reduce tartrate crystals (generally potassium bitartrate) in wine. These tartrate crystals look like grains of clear sand, and are also known as "wine crystals" or "wine diamonds". They are formed by the union of tartaric acid and potassium, and may appear to be sediment in the wine, though they are not. During the cold stabilizing process after fermentation, the temperature of the wine is dropped to close to freezing for 1–2 weeks. This will cause the crystals to separate from the wine and stick to the sides of the holding vessel. When the wine is drained from the vessels, the tartrates are left behind. They may also form in wine bottles that have been stored under very cold conditions.

During "heat stabilization", unstable proteins are removed by adsorption onto bentonite, preventing them from precipitating in the bottled wine.

Secondary Fermentation and Bulk Aging

During the secondary fermentation and aging process, which takes three to six months, the fermentation continues very slowly. The wine is kept under an airlock to protect the wine from oxidation. Proteins from the grape are broken down and the remaining yeast cells and other fine particles from the grapes are allowed to settle. Potassium bitartrate will also precipitate, a process which can be enhanced by cold stabilization to prevent the appearance of (harmless) tartrate crystals after bottling. The result of these processes is that the originally cloudy wine becomes clear. The wine can be racked during this process to remove the lees.

The secondary fermentation usually takes place in either large stainless steel vessels with a volume of several cubic meters, or oak barrels, depending on the goals of the winemakers. Unoaked wine is fermented in a barrel made of stainless steel or other material having no influence in the final taste of the wine. Depending on the desired taste, it could be fermented mainly in stainless steel to be briefly put in oak, or have the complete fermentation done in stainless steel. Oak could be added as chips used with a non-wooden barrel instead of a fully wooden barrel. This process is mainly used in cheaper wine.

Amateur winemakers often use glass carboys in the production of their wine; these vessels (sometimes called *demijohns*) have a capacity of 4.5 to 54 litres (1.2–14.3 US gallons). The kind of vessel used depends on the amount of wine that is being made, the grapes being used, and the intentions of the winemaker.

Malolactic Fermentation

Malolactic fermentation occurs when lactic acid bacteria metabolize malic acid and produce lactic acid and carbon dioxide. This is carried out either as an intentional procedure in which specially cultivated strains of such bacteria are introduced into the maturing wine, or it can happen by chance if uncultivated lactic acid bacteria are present.

Malolactic fermentation can improve the taste of wine that has high levels of malic acid, because malic acid in higher concentration generally causes an often unpleasant harsh and bitter taste sensation, whereas lactic acid is perceived as more gentle and less sour.

The process is used in most red wines and is discretionary for white wines.

Laboratory Tests

Whether the wine is aging in tanks or barrels, tests are run periodically in a laboratory to check the status of the wine. Common tests include °Brix, pH, titratable acidity, residual sugar, free or available sulfur, total sulfur, volatile acidity and percent alcohol. Additional tests include those for the crystallization of cream of tartar (potassium hydrogen tratrate) and the precipation of heat unstable protein; this last test is limited to white wines. These tests are often performed throughout the making of the wine as well as prior to bottling. In response to the results of these tests, a winemaker can then decide on appropriate remedial action, for example the addition of more sulfur dioxide. Sensory tests will also be performed and again in response to these a wine maker may take remedial action such as the addition of a protein to soften the taste of the wine.

Brix is one measure of the soluble solids in the grape juice and represents not only the sugars but also includes many other soluble substances such as salts, acids and tannins, sometimes called Total Soluble Solids (TSS). However, sugar is by far the compound in greatest quantity and so for all practical purposes these units are a measure of sugar level. The level of sugar in the grapes is important not only because it will determine the final alcohol content of the wine, but also because it is an indirect index of grape maturity. Brix (Bx for short) is measured in grams per hundred grams of solution, so 20 Bx means that 100 grams of juice contains 20gm of dissolved compounds. There are other common measures of sugar content of grapes, Specific gravity, Oechsle (Germany) and Beaume (France). The French Baume has the benefit that one Be° gives approximately one percent alcohol. Also one Be° is equal to 1.8 Brix, that is 1.8 grams of sugar per one hundred grams. This helps with deciding how much sugar to add if the juice is low in sugar: to achieve one percent alcohol add 1.8 grams per 100 ml or 18 grams per liter. This process is called chaptalization and is illegal in some countries (but perfectly acceptable for the home winemaker.) Generally, for the making of dry table wines a Bx of between 20 and 25 is desirable (equivalent to Be° of 11 to 14.)

A Brix test can be run either in the lab or in the field for a quick reference number to see what the sugar content is. Brix is usually measured with a refractometer while the other methods use a hydrometer. Generally, hydrometers are a cheaper alternative. For more accurate use of sugar measurement it should be remembered that all measurements are affected by the temperature at which the reading is made. Suppliers of equipment generally will supply correction charts.

Volatile acidity test verifies if there is any steam distillable acids in the wine. Mainly present is acetic acid but lactic, butyric, propionic and formic acids can also be found. Usually the test checks for these acids in a cash still, but there are new methods available such as HPLC, gas chromatography and enzymatic methods. The amount of volatile acidity found in sound grapes is negligible, since it is a by-product of microbial metabolism. It's important to remember that acetic acid bacteria require oxygen to grow. Eliminating any air in wine containers as well as a sulfur dioxide addition will limit their growth. Rejecting moldy grapes will also prevent possible problems associated with acetic acid bacteria. Use of sulfur dioxide and inoculation with a low-V.A. producing strain of

Saccharomyces may deter acetic acid producing yeast. A relatively new method for removal of volatile acidity from a wine is reverse osmosis. Blending may also help—a wine with high V.A. can be filtered (to remove the microbe responsible) and blended with a low V.A. wine, so that the acetic acid level is below the sensory threshold.

Blending and Fining

Different batches of wine can be mixed before bottling in order to achieve the desired taste. The winemaker can correct perceived inadequacies by mixing wines from different grapes and batches that were produced under different conditions. These adjustments can be as simple as adjusting acid or tannin levels, to as complex as blending different varieties or vintages to achieve a consistent taste.

Fining agents are used during winemaking to remove tannins, reduce astringency and remove microscopic particles that could cloud the wines. The winemakers decide on which fining agents are used and these may vary from product to product and even batch to batch (usually depending on the grapes of that particular year).

Gelatin has been used in winemaking for centuries and is recognized as a traditional method for wine fining, or clarifying. It is also the most commonly used agent to reduce the tannin content. Generally no gelatin remains in the wine because it reacts with the wine components, as it clarifies, and forms a sediment which is removed by filtration prior to bottling.

Besides gelatin, other fining agents for wine are often derived from animal and fish products, such as micronized potassium casseinate (casein is milk protein), egg whites, egg albumin, bone char, bull's blood, isinglass (Sturgeon bladder), PVPP (a synthetic compound), lysozyme, and skim milk powder.

Some aromatized wines contain honey or egg-yolk extract.

Non-animal-based filtering agents are also often used, such as bentonite (a volcanic clay-based filter), diatomaceous earth, cellulose pads, paper filters and membrane filters (thin films of plastic polymer material having uniformly sized holes).

Fermentation-Turning Grape Juice into Alcohol

Grape juice is turned into alcohol by the process of "fermentation." Grapes on the vine are covered with yeast, mold and bacteria. By putting grape juice into a container at the right temperature, yeast will turn the sugar in the juice into alcohol and carbon dioxide. The grape juice will have fermented. Fermentation is carried out in stainless steel vessels.

Yeast also gives flavor to the wine. But the yeast that is on the grape skin when it is harvested may not have the desired flavor. Other things on the outside of a grape are not good for wine (for example, acetic bacteria on the grapes can cause the wine to turn to vinegar). The winemaker can eliminate unwanted yeast's, molds and bacteria, most commonly by using the "universal disinfectant," sulfur dioxide. Unfortunately, the sulfites which remain in the wine may cause a lot of discomfort to some wine drinkers. Some winemakers prefer NOT to do this, and purposely create wines that are subject to the vagaries (and different flavours) of the yeast that pre-exist on the grapes ("wild yeast fermentation").

The winemaker has many different yeast strains to choose from (and can use different strains at different times during the process for better control fermentation). The most common wine yeast is Saccharomyces.

This is a good point to stop and mention "Brett," also known as the Brettanomyces strain of yeast (which can be added or come from wild yeast fermentation). As yeast works, it causes grape juice ("must") to get hot. But if there's too much heat, the yeast won't work. Cooling coils are necessary to maintain a temperature below 45 C.

A less modern, but still wide widely used way to ferment wine is to place it in small oak barrels. "Barrel fermentation" is usually done at a lower temperature in temperature controlled rooms and takes longer, perhaps around 6 weeks. The longer fermentation and use of wood contributes to the flavour (and usually expense) of the wine.

The skins and pulp which remain in a red wine vat will rise to and float on top of the juice. This causes problems (if it dries out, it's a perfect breeding ground for injurious bacteria), so the winemaker will push this "cap" back down into the juice, usually at least twice a day. In large vats, this is accomplished by pumping juice from the bottom of the vat over the top of the cap.

Eventually the yeast is no longer changing sugar to alcohol (though different strains of yeast, which can survive in higher and higher levels of alcohol, can take over and contribute their own flavour to the wine-as well as converting a bit more sugar to alcohol).

After all this is completed what you have left is the wine, "dead" yeast cells, known as "lees and various other substances.

Malo-lactic Fermentation

The winemaker may choose to allow a white wine to undergo a second fermentation which occurs due to malic acid in the grape juice. When malic acid is allowed to break down into carbon dioxide and lactic acid (thanks to bacteria in the wine), it is known as "malo-lactic fermentation," which imparts additional flavour to the wine. A "buttery" flavour in some whites is due to this process. This process is used for sparkling wines.

First Racking

After fermentation completed naturally or stopped by addition of distilled spirit, first racking is carried out. This involves the wine to stand still until most yeast cells and fine suspended material settle out. The wine is then filtered without disturbing the sediment or the yeast.

Winery Aging

The winery may then keep the wine so that there can be additional clarification and, in some wines, to give it a more complex flavours. Flavour can come from wood (or more correctly from the chemicals that make up the wood and are taken up into the wine).

The wine may be barrel aged for several months to several years. No air is allowed to enter the barrels during this period.

Ignoring any additional processing that might be used, you could empty the barrels into bottles and sell your wine. However, during the winery aging, the smaller containers may develop differences. So the winemaker will probably "blend" wine from different barrels, to achieve a uniform result. Also, the winemaker may blend together different grape varieties to achieve desired characteristics.

Stabilization, Filtration

Stabilization is carried out to remove traces of tartaric acid. These tartarates present in the grape juice tend to crystallize in wine and if not removed completely can slowly reappear as glass like crystals in final bottles on storage.

Stabilization with respect to tartarates may involve chilling of wine that can crystallize tartarates and these crystals can be removed by filtration.

Pasteurization

If the wine has an alcohol content less than 14% it may be heat pasteurized or cold pasteurized through microporous filters just before bottling.

Preservatives

The most common preservative used in winemaking is sulfur dioxide, achieved by adding sodium or potassium metabisulphite. Another useful preservative is potassium sorbate.

Sulfur dioxide has two primary actions, firstly it is an anti microbial agent and secondly an anti oxidant. In the making of white wine it can be added prior to fermentation and immediately after alcoholic fermentation is complete. If added after alcoholic ferment it will have the effect of preventing or stopping malolactic fermentation, bacterial spoilage and help protect against the damaging effects of oxygen. Additions of up to 100 mg per liter (of sulfur dioxide) can be added, but the available or free sulfur dioxide should be measured by the aspiration method and adjusted to 30 mg per liter. Available sulfur dioxide should be maintained at this level until bottling. For rose wines smaller additions should be made and the available level should be no more than 30 mg per liter.

In the making of red wine sulfur dioxide may be used at high levels (100 mg per liter) prior to ferment to assist stabilize colour otherwise it is used at the end of malolactic ferment and performs the same functions as in white wine. However, small additions (say 20 mg per liter) should be used to avoid bleaching red pigments and the maintenance level should be about 20 mg per liter. Furthermore, small additions (say 20 mg per liter) may be made to red wine after alcoholic ferment and before malolactic ferment to overcome minor oxidation and prevent the growth of acetic acid bacteria.

Without the use of sulfur dioxide, wines can readily suffer bacterial spoilage no matter how hygienic the winemaking practice.

Potassium sorbate is effective for the control of fungal growth, including yeast, especially for sweet wines in bottle. However, one potential hazard is the metabolism of sorbate to geraniol a potent and very unpleasant by-product. To avoid this, either the wine must be sterile bottled or contain enough sulfur dioxide to inhibit the growth of bacteria. Sterile bottling includes the use of filtration.

Filtration

Filtration in winemaking is used to accomplish two objectives, clarification and microbial stabilization. In clarification, large particles that affect the visual appearance of the wine are removed. In microbial stabilization, organisms that affect the stability of the wine are removed therefore reducing the likelihood of re-fermentation or spoilage.

The process of clarification is concerned with the removal of particles; those larger than 5–10 micrometers for coarse polishing, particles larger than 1–4 micrometers for clarifying or polishing. Microbial stabilization requires a filtration of at least 0.65 micrometers. However, filtration at this level may lighten a wines colour and body. Microbial stabilization does not imply sterility. It simply means that a significant amount of yeast and bacteria have been removed.

Bottling Wine

Producers often use different shaped bottles to denote different types of wine. Coloured bottles help to reduce damage by light. (Light assists in oxidation and breakdown of the wine into chemicals, such as mercaptan, which are undesirable.)

Cellaring Wine

Most people assume that the longer that you keep a wine, the better it will get Since its best to store wine under certain conditions, like in a cool damp underground cellar, this is known as "cellaring" wine.

It is a misconception that you MUST age wine. The fact is, throughout the world, most wine is drunk "young" (that is relatively soon after it is produced, perhaps 12 to 18 months), even wines that are "better" if aged. While some wines will "mature" and become better over time, others will not and should be drunk immediately, or within a few years.

Tannin is a substance that comes from the seeds, stems and skins of grapes. Additional tannin can come from the wood during barrel aging in the winery. It is a preservative and is important to the long term maturing of wine. Through time, tannin (which has a bitter flavour) will precipitate out of the wine (becoming sediment in the bottle) and the complexity of the wine's flavour from fruit, acid and all the myriad other substances that make up the wine's character will come into greater balance. Generally, it is red wines that are the ones that CAN (but do not have to be) produced with a fair amount of tannin with an eye towards long term storing and maturation. The bad news is that you shouldn't drink it young since it will taste too harsh (and probably cost too much, besides). The good news is that after a number of years, what you get is a prized, complex and balanced wine.

Remember that red wines get their colour from the stems and skins of the grape. This gives the wine tannin and aging capacity. White wines may have no contact with the stems and skins and will have little tannin (though some can be added, again, through barrel aging). Therefore most white wines don't age well. Even the ones which do get better through time will not last nearly as long as their red cousins. A fair average for many "ageable" whites would be about 5 to 7 years (some might go 10). On the other hand, really "ageable" reds can easily be kept for 30 years and longer.

Storing Wine

For wines that should be aged, a cellar should have proper :Temperature which does not have rapid fluctuation. 55 degrees Fahrenheit is a good, but you can live with 50 to 57 degrees Fahrenheit (10 to 14 degrees Centigrade). Wide swings in temperature will harm the wine. Having too high a temperature will age the wine faster so it won't get as complex as it might have. Having too low a temperature will slow the wine's maturation. Humidity. About 60 percent is right. This helps keep the cork moist. The wine will oxidize if the air (and its oxygen) gets to it. If the cork dries out, it can shrink and let air in. This is another reason to keep the bottles on their sides. The wine itself will help keep the cork moist.

Pectic Enzymes on Clarification of Wines

The enzymes such as pectinase that hydrolyze the large pectin molecules.

Pectinase

An enzyme that catalyses the hydrolysis of pectin molecules.

Pectin

A heavy, colloidal substance found in most ripe fruit which promotes the formation of gelatinous solutions and hazes in the finished wine. Fermenting fruit pulps with high pectin content, such as apples, should be treated with pectic enzyme, especially if the pulp is boiled to extract the fruit flavour (boiling releases the pectin, while pectic enzymes destroy it). Pectin is a structural heteropolysaccharide contained in the primary cell walls of terrestrial plants. It was first isolated and described in 1825 by Henri Braconnot. It is produced commercially as a white to light brown powder, mainly extracted from citrus fruits, and is used in food as a gelling agent particularly in jams and jellies. It is also used in fillings, sweets, as a stabilizer in fruit juices and milk drinks and as a source of dietary fibre.

Biology

In plant cells, pectin consists of a complex set of polysaccharides that are present in most primary cell walls and particularly abundant in the non-woody parts of terrestrial plants. Pectin is present not only throughout primary cell walls but also in the middle lamella between plant cells where it helps to bind cells together.

The amount, structure and chemical composition of pectin differs between plants, within a plant over time and in different parts of a plant. During ripening, pectin is broken down by the enzymes pectinase and pectinesterase; in this process the fruit becomes softer as the middle lamella breaks down and cells become separated from each other. A similar process of cell separation caused by pectin breakdown occurs in the abscission zone of the petioles of deciduous plants at leaf fall.

Pectin is a natural part of human diet, but does not contribute significantly to nutrition. The daily intake of pectin from fruits and vegetables can be estimated to be around 5 g (assuming consumption of approximately 500 g fruits and vegetables per day).

In human digestion, pectin goes through the small intestine more or less intact. Pectin is thus a soluble dietary fibre.

Consumption of pectin has been shown to reduce blood cholesterol levels. The mechanism appears to be an increase of viscosity in the intestinal tract, leading to a reduced absorption of cholesterol from bile or food. In the large intestine and colon, microorganisms degrade pectin and liberate short-chain fatty acids that have positive influence on health (prebiotic effect).

Chemistry

Pectins are a family of complex polysaccharides that contain 1,4-linked α-D-galactosyluronic acid residues. Three pectic polysaccharides have been isolated from plant primary cell walls and structurally characterized. These are:

- Homogalacturonans
- Substituted galacturonans
- Rhamnogalacturonans.

Homogalacturonans are linear chains of α-(1-4)-linked D-galacturonic acid.

Substituted galacturonans are characterized by the presence of saccharide appendant residues (such as D-xylose or D-apiose in the respective cases of xylogalacturonan and apiogalacturonan) branching from a backbone of D-galacturonic acid residues.

Rhamnogalacturonan I pectins (RG-I) contain a backbone of the repeating disaccharide: 4)-α-D-galacturonic acid-(1,2)-α-L-rhamnose-(1. From many of the rhamnose residues, sidechains of various neutral sugars branch off. The neutral sugars are mainly D-galactose, L-arabinose and D-xylose, the types and proportions of neutral sugars varying with the origin of pectin. Another structural type of pectin is rhamnogalacturonan II (RG-II), which is a less frequent complex, highly branched polysaccharide. Rhamnogalacturonan II is classified by some authors within the group of substituted galacturonans since the rhamnogalacturonan II backbone is made exclusively of D-galacturonic acid units.

Isolated pectin has a molecular weight of typically 60–130,000 g/mol, varying with origin and extraction conditions.

In nature, around 80% of carboxyl groups of galacturonic acid are esterified with methanol. This proportion is decreased more or less during pectin extraction. The ratio of esterified to non-esterified galacturonic acid determines the Behaviour of pectin in food applications. This is why pectins are classified as high- vs. low-ester pectins – or in short HM vs. LM-pectins, with more or less than half of all the galacturonic acid esterified.

The non-esterified galacturonic acid units can be either free acids (carboxyl groups) or salts with sodium, potassium or calcium. The salts of partially esterified pectins are called pectinates, if the degree of esterification is below 5% the salts are called pectates, the insoluble acid form, pectic acid.

Some plants like sugar beet, potatoes and pears contain pectins with acetylated galacturonic acid in addition to methyl esters. Acetylation prevents gel-formation but increases the stabilising and emulsifying effects of pectin.

Amidated pectin is a modified form of pectin. Here, some of the galacturonic acid is converted with ammonia to carboxylic acid amide. These pectins are more tolerant of varying calcium concentrations that occur in use. To prepare a pectin-gel, the ingredients are heated, dissolving the pectin. Upon cooling below gelling temperature, a gel starts to form. If gel formation is too strong, syneresis or a granular texture are the result, whilst weak gelling leads to excessively soft gels. In high-ester pectins at soluble solids content above 60% and a pH-value between 2.8 and 3.6, hydrogen bonds and hydrophobic interactions bind the individual pectin chains together. These bonds form as water is bound by sugar and forces pectin strands to stick together. These form a 3-dimensional molecular net that creates the macromolecular gel. The gelling-mechanism is called a low-water-activity gel or sugar-acid-pectin gel.

In low-ester pectins, ionic bridges are formed between calcium ions and the ionised carboxyl groups of the galacturonic acid. This is idealised in the so-called "egg box-model". Low-ester pectins need calcium to form a gel, but can do so at lower soluble solids and higher pH-values than high-ester pectins. Amidated pectins behave like low-ester pectins but need less calcium and are more tolerant of excess calcium. Also, gels from amidated pectin are thermo-reversible – they can be heated and after cooling solidify again, whereas conventional pectin-gels will afterwards remain liquid.

High-ester pectins set at higher temperatures than low-ester pectins. However, gelling reactions with calcium increase as the degree of esterification falls. Similarly, lower pH-values or higher soluble solids (normally sugars) increase gelling speed. Suitable pectins can therefore be selected for jams and for jellies, or for higher sugar confectionery jellies.

Sources and Production

Apples, guavas, quince, plums, gooseberries, oranges and other citrus fruits, contain large amounts of pectin, while soft fruits like cherries, grapes and strawberries contain small amounts of pectin.

Typical levels of pectin in plants are (fresh weight):

- apples, 1–1.5%
- apricot, 1%

- cherries, 0.4%
- oranges 0.5–3.5%
- carrots approx. 1.4%
- citrus peels, 30%.

The main raw-materials for pectin production are dried citrus peel or apple pomace, both by-products of juice production. Pomace from sugar-beet is also used to a small extent.

From these materials, pectin is extracted by adding hot dilute acid at pH-values from 1.5 – 3.5. During several hours of extraction, the protopectin loses some of its branching and chain-length and goes into solution. After filtering, the extract is concentrated in vacuum and the pectin then precipitated by adding ethanol or isopropanol. An old technique of precipitating pectin with aluminium salts is no longer used (apart from alcohols and polyvalent cations; pectin also precipitates with proteins and detergents).

Alcohol-precipitated pectin is then separated, washed and dried. Treating the initial pectin with dilute acid leads to low-esterified pectins. When this process includes ammonium hydroxide, amidated pectins are obtained. After drying and milling pectin is usually standardised with sugar and sometimes calcium-salts or organic acids to have optimum performance in a particular application.

Worldwide, approximately 40,000 metric tons of pectin are produced every year.

Uses

The main use for pectin is as a gelling agent, thickening agent and stabilizer in food. The classical application is giving the jelly-like consistency to jams or marmalades, which would otherwise be sweet juices. For household use, pectin is an ingredient in gelling sugar (also known as "Jam Sugar") where it is diluted to the right concentration with sugar and some citric acid to adjust pH. In some countries, pectin is also available as a solution or an extract, or as a blended powder, for home jam making. For conventional jams and marmalades that contain above 60% sugar and soluble fruit solids, high-ester pectins are used. With low-ester pectins and amidated pectins less sugar is needed, so that diet products can be made. Pectin can also be used to stabilize acidic protein drinks, such as drinking yogurt, and as a fat substitute in baked goods. Typical levels of pectin used as a food additive are between 0.5 – 1.0% - this is about the same amount of pectin as in fresh fruit.

In medicine, pectin increases viscosity and volume of stool so that it is used against constipation and diarrhea. Until 2002, it was one of the main ingredients used in Kaopectate, along with kaolinite. Pectin is also used in throat lozenges as a demulcent. In cosmetic products, pectin acts as stabilizer. Pectin is also used in wound healing preparations and speciality medical adhesives, such as colostomy devices. Also, it is considered a natural remedy for nausea. Pectin rich foods are proven to help nausea.

In ruminant nutrition, depending on the extent of lignification of the cell wall, pectin is up to 90% digestible by bacterial enzymes. Ruminant nutritionists recommend that the digestibility and energy concentration in forages can be improved by increasing pectin concentration in the forage.

In the cigar industry, pectin is considered an excellent substitute for vegetable glue and many cigar smokers and collectors will use pectin for repairing damaged tobacco wrapper leaves on their cigars.

Pectin is also used in jellybeans.

The Effect of Pectic Enzymes in Wine Making

What are Pectins? How do they Influence the Clarification of Wine?

Broadly speaking, pectins constitute a group of closely related carbohydrates allied to such polysaccharides as starch and chemically defined as complex, methylated polygalacturonic acids. They possess colloidal properties and will gel (i.e. set to jelly in the presence of high concentrations of sugar and acid). The essential role of pectins as setting agents in jams and jellies has long been recognized. For making jam and jelly fruits which are naturally high in pectin are a decided advantage. Exactly the opposite is true in winemaking where even a small amount of pectin can cause many problems.

Although winemakers rarely encounter conditions which cause musts and wines high in pectin to gel, the fact that some is present can have other equally undesirable effects. Thus, wines containing pectin are exceptionally difficult to clear because the pectin acts as a colloid and stabilizes the haze. The minute particles which are responsible for the wine appearing cloudy are too small to settle out naturally no matter how long a period of storage is allowed. Wines only clear because the tiny suspended particles constituting the haze are mutually attracted and combined to form larger particles whose size eventually increases sufficiently for them to begin settling under the force of gravity. The action of protective colloid such as pectin and starch is to oppose these processes of clarification by stabilizing the haze forming particles and preventing their agglomeration. Since there is no tendency for the particles to attract each other, there is no particle growth and hence no settling so the wine remains stubbornly cloudy indefinitely. Pectin stabilizes haze so efficiently that neither fining nor filtering is likely to be effective and may only make matters worse.

How, then, can wines whose haze forming particles are stabilized by pectin be clarified? The only satisfactory way to clear pectin clouded wines is to degrade the pectin so that the protective action on the particles causing the haze is destroyed and agglomeration can occur in the normal manner. This is accomplished by the addition of ½ teaspoon of Pectic Enzyme Powder per gallon of wine. The essential function of these enzymes is to break the long pectin chain into shorter units which are unable to act as protective colloids.

Although the addition of Pectic Enzymes can be used with great success for clearing wines with pectin stabilized hazes, it is obviously preferable to avoid encountering such difficulties at such a late stage in the production of the wine. Far less trouble would occur if the pectin could be destroyed prior to and/or during fermentation so that it is completely degraded by the time the first rack occurs. The wine would then clear more quickly and more completely. So for best results, add ½ teaspoon of Pectic Enzyme Powder to all your wines prior to the addition of yeast.

When fruit is added to beer or mead, it should also be treated with Pectic Enzymes to prevent pectin stabilized hazes.

1. Yields of free run and total yields of wine per ton of grapes have on the average been greater from enzyme treated than from untreated crushed grapes.
2. Use of pectic enzymes in crushed white grapes almost invariably deepened the colour of the wine in comparison with that of wine made from grapes to which no enzyme was added—probably owing to greater extraction of colour from the skins.
3. Colour of wines made from free run juice to which was added a pectic enzyme was in mamy cases about like that of the untreated wine, but also in about an equal number of cases the colour of the treated wines on aging became slightly deeper (yellower) than that of the untreated wines.
4. Therefore, if light colour is the primary objective, the enzyme should be added to the white juice rather than to the crushed grapes.
5. Enzyme-treated wines have invariably matured more rapidly than the untreated. The wine maker must be careful not to let the treated wines overmature before bottling. In one winery experiment of the 1950 season the enzyme-treated wine is now inferior to the untreated in flavour and bouquet. In other cases there is little difference, or the enzyme treated is superior.
6. Enzyme-treated wines developed much more compact lees than the treated and were racked with less danger of carrying lees over into the wine.

7. As in previous seasons (1936-1949), we found that the enzyme-treated wines cleared much more rapidly than the untreated.
8. Also, they filtered much more rapidly and much more wine could be filtered with each setting of the filter.
9. Much less bentonite was needed for clarification of the enzyme-treated wines—usually 1-2 pounds per 1000 gallons as against 4-10 pounds for the untreated.
10. At two winners relatively small amounts of Pectineal have been used to pound of Pectinol-0 per ton. The wines so treated do not clear as rapidly as with larger additions of enzyme but are found to filter very much more rapidly than the untreated and to require far less bentonite in clarification.
11. Enzyme-treated juices were found commercially to filter more readily and to give concentrates of better quality than the untreated.
12. As reported by Berg and Marsh, it was observed that Pectineal greatly increased the rate of clearing of heat extracted fermented red musts. Usually the enzyme-treated heat extracted wines in our experiments were of deeper colour than the checks, indicating that the enzyme may have stabilized the colour.
13. On the other hand, where the enzyme was used in the crushed grapes without heat extraction the colour of the resulting dry wines were in some cases less intense than in the check lots. In others it was more intense or about equal to the untreated.
14. No very clear cut differences were observed in the effect of several different Rohm and Haas enzyme preparations on the colour of either red or white wines.
15. Enzyme-treated white wines darkened more rapidly than the untreated when incubated at 45° C (131° F), although the difference was usually not very great.
16. Preliminary evidence was found to indicate that galacturonic acid formed by hydrolysis of the pectin of the must by the enzyme may be concerned in the darkening of the colour of enzyme-treated wines.
17. Pectineal solutions of the concentrations applied in wine making did not darken; indicating that where darkening of the treated wine occurs it is some constituent of the wine itself that is concerned.
18. Ports made with Pectineal were in most cases deeper in colour than the untreated.
19. Enzyme-treated musts foamed much less than the untreated during fermentationo

20. There appeared to be no consistent differences in heat and cold stability of enzyme-treated and untreated wines. Finished enzyme-treated dry white wines were stable at 131° F and at 0° F.
21. Maintenance of a reasonable level of SO_2 in enzyme-treated white wines held the colour.

Vinegar Fermentation and Production

Vinegar is an acid liquid produced from the fermentation of ethanol in a process that yields its key ingredient, acetic acid (ethanoic acid). It also may come in a diluted form. The acetic acid concentration typically ranges from 4% to 8% by volume for table vinegar and up to 18% for pickling. Natural vinegars also contain small amounts of tartaric acid, citric acid, and other acids. Vinegar has been used since ancient times and is an important element in European, Asian, and other cuisines.

The word "vinegar" derives from the Old French *vin aigre,* meaning "sour wine", which in turn is derived from the Latin "vinum aegrum" meaning "feeble wine".

History

Vinegar has been made and used for thousands of years. Traces of it have been found in Egyptian urns dating from around 3000 BC. According to Shennong's *Herb Classic,* vinegar was invented in China during the Xia Dynasty, around 2000 BC.

In the Tanakh and the Old Testament of the Bible, it is mentioned as unpleasant to drink (Ps. 69:21) and foolish to combine with *nether* (most likely soda ash, although possibly potash, natron, or niter) (Prov. 25:20), but more favorably as a condiment when Boaz allows Ruth to "dip her piece of bread in the vinegar" (Ruth 2:14). Jesus was offered vinegar or sour wine while on the cross (Matthew 27:48; Mark 15:36). In Islamic traditions, vinegar is one of the four favoured condiments of the Prophet Muhammad, who called it a "Blessed seasoning".

In 1864, Louis Pasteur showed that vinegar results from a natural fermentation process.

Production

Vinegar is made from the fermentation of ethanol by acetic acid bacteria. The ethanol may be derived from many different sources including wine, cider, beer or fermented fruit juice, or it may be made synthetically from natural gas and petroleum derivatives.

Commercial vinegar is produced either by fast or slow fermentation processes. Slow methods generally are used with traditional vinegars and fermentation proceeds slowly over the course of weeks or months. The longer fermentation period allows for the accumulation of a nontoxic slime composed of acetic acid bacteria and soluble cellulose, known as the mother of vinegar. Fast methods add mother of vinegar (i.e., bacterial culture) to the source liquid before adding air using a Venturi pump system or a turbine to promote oxygenation to obtain the fastest fermentation. In fast production processes, vinegar may be produced in a period ranging from 20 hours to three days.

Varieties

Malt

Malt vinegar is made by malting barley, causing the starch in the grain to turn to maltose. Then an ale is brewed from the maltose and allowed to turn into vinegar, which is then aged. It typically is light brown in colour.

Wine

Wine vinegar is made from red or white wine and is the most commonly used vinegar in Mediterranean countries and Central Europe. As with wine, there is a considerable range in quality. Better quality wine vinegars are matured in wood for up to two years and exhibit a complex, mellow flavour. Wine vinegar tends to have a lower acidity than that of white or cider vinegars. There are more expensive wine vinegars that are made from individual varieties of wine, such as Champagne, Sherry, or pinot grigio.

Apple Cider

Apple cider vinegar, otherwise known simply as cider vinegar or ACV, is made from cider or apple must and has a brownish-yellow colour. It often is sold unfiltered and unpasteurized with the mother of vinegar present, as a natural product. It is very popular, partly because of supposed beneficial health and beauty properties. Because of its acidity, apple cider vinegar may be very harsh, even burning, to the throat. If taken straight, (as opposed to used in cooking), it can be diluted (e.g., with fruit juice or water) before drinking. It is also sometimes sweetened with sugar or honey. There have been reports of acid chemical burns of the throat from apple cider vinegar tablets, but "doubt remains as to whether apple cider vinegar was in fact an ingredient in the evaluated products."

Fruit

Fruit vinegars are made from fruit wines, usually without any additional flavouring. Common flavours of fruit vinegar include apple, black currant, raspberry, quince, and tomato. Typically, the flavours of the original fruits remain in the final product.

Most fruit vinegars are produced in Europe, where there is a growing market for high-priced vinegars made solely from specific fruits (as opposed to non-fruit vinegars which are infused with fruits or fruit flavours). Several varieties, however, also are produced in Asia. Persimmon vinegar, called *gam sikcho*, is popular in South Korea. Jujube vinegar and wolfberry vinegar are produced in China.

Umezu, a salty, sour liquid that is a by-product of *umeboshi* (pickled *ume*) production, is produced in Japan, but technically is not a true vinegar.

Jamun Sirka is a vinegar produced from the *Jamun* (or rose apple) fruit in India. It is considered to be medicinally valuable for stomach, spleen and diabetic ailments.

Balsamic

Balsamic vinegar is an aromatic, aged type of vinegar traditionally crafted in the Modena and Reggio Emilia provinces of Italy from the concentrated juice, or must, of white grapes (typically of the Trebbiano variety). It is very dark brown in colour and its flavour is rich, sweet, and complex, with the finest grades being the product of years of aging in a successive number of casks made of various types of wood (including oak, mulberry, chestnut, cherry, juniper, ash, and acacia). Originally a product available only to the Italian upper classes, a cheaper form of balsamic vinegar became widely known and available around the world in the late twentieth century.

True balsamic vinegar (which has Protected Designation of Origin) is aged for 12 to 25 years. Balsamic vinegars that have been aged for up to 100 years are available, though they are usually very expensive. The commercial balsamic sold in supermarkets is typically made with concentrated grape juice mixed with a strong vinegar, which is laced with caramel and sugar. Regardless of how it is produced, balsamic vinegar must be made from a grape product.

Balsamic vinegar has a high acidity level but the tart flavour is usually hidden by the sweetness of the other ingredients, making it very mellow.

Rice

Rice vinegar is most popular in the cuisines of East and Southeast Asia. It is available in "white" (light yellow), red, and black varieties. The Japanese prefer a light rice vinegar for the preparation of sushi rice and salad dressings. Red rice vinegar traditionally is coloured with red yeast rice. Black rice vinegar (made with black glutinous rice) is most popular in China, and it is also widely used in other east Asian countries.

White rice vinegar has a mild acidity and a somewhat "flat", uncomplex flavour. Some varieties of rice vinegar are sweetened or otherwise seasoned with spices or other added flavorings.

Coconut

Coconut vinegar, made from fermented coconut water, is used extensively in Southeast Asian cuisine (particularly in the Philippines, a major producer, where it is called *suka ng niyog*), as well as in some cuisines of India. A cloudy white liquid, it has a particularly sharp, acidic taste with a slightly yeasty note.

Palm

Palm vinegar, made from the fermented sap from flower clusters of the nipa palm (also called attap palm), is used most often in the Philippines, where it is produced, and where it is called *sukang paombong.*

Cane

Cane vinegar, made from sugar cane juice, is most popular in the Philippines, in particular, the Ilocos Region of the northern Philippines (where it is called *sukang iloko*), although it also is produced in France and the United States. It ranges from dark yellow to golden brown in colour and has a mellow flavour, similar in some respects, to rice vinegar, though with a somewhat "fresher" taste. Contrary to expectation, containing no residual sugar, it is not sweeter than other vinegars. In the Philippines, it often is labelled as *suka na maasim*, although this is simply a generic term meaning "sour vinegar."

Cane vinegars from Ilocos also varies in two different types. The "basi" or the sweet vinegar and the "Suka" which is the sour one. The sweet vinegar is used as a wine in Ilokanos, while the other type of vinegar is used as a seasoning and preservative. A white variation has become quite popular in Brazil in recent years, where it is the cheapest type of vinegar sold. It is now common for other types of vinegar (made from wine, rice and apple cider) to be sold mixed with cane vinegar to lower the costs.

Raisin

Vinegar made from raisins, called *khal 'anab* is used in cuisines of the Middle East, and is produced therein. It is cloudy and medium brown in colour, with a mild flavour.

Date

Vinegar made from dates is a traditional product of the Middle East.

Beer

Vinegar made from beer is produced in the United Kingdom, Germany,

Austria, and the Netherlands. Although its flavour depends on the particular type of beer from which it is made, it is often described as having a malty taste. That produced in Bavaria, is a light golden colour with a very sharp and not-overly-complex flavour.

Honey

Vinegar made from honey is rare, although commercially available honey vinegars are produced in Italy, France, Romania and Spain.

East Asian Black

Chinese black vinegar is an aged product made from rice, wheat, millet, sorghum, or a combination thereof. It has an inky black colour and a complex, malty flavour. There is no fixed recipe and thus some Chinese black vinegars may contain added sugar, spices, or caramel colour. The most popular variety, Zhenjiang vinegar, originated in the city of Zhenjiang, in the eastern coastal province of Jiangsu, China and also is produced in Tianjin and Hong Kong.

A somewhat lighter form of black vinegar, made from rice, also is produced in Japan, where it is called *kurozu.* Since 2004 it has been marketed as a healthful drink; its manufacturers claim that it contains high concentrations of amino acids.

Flavored Vinegars

Popular fruit-flavoured vinegars include those infused with whole raspberries, blueberries, or figs (or else from flavourings derived from these fruits). Some of the more exotic fruit-flavoured vinegars include blood orange and pear.

Herb vinegars are flavoured with herbs, most commonly Mediterranean herbs such as thyme or oregano. Such vinegars can be prepared at home by adding sprigs of fresh or dried herbs to vinegar purchased at a grocery store; generally a light-coloured, mild tasting vinegar, such as that made from white wine, is used for this purpose.

Sweetened vinegar is of Cantonese origin and is made from rice wine, sugar and herbs including ginger, cloves, and other spices.

Job's Tears

In Japan, an aged vinegar also is made from Job's Tears, a tall grain-bearing tropical plant. The vinegar is similar in flavour to rice vinegar.

Kombucha

Kombucha vinegar is made from kombucha, a symbiotic culture of yeast and bacteria. The bacteria produce a complex array of nutrients and populate the vinegar with bacteria which some claim promotes a healthy digestive tract, although no scientific studies have confirmed this. Kombucha vinegar primarily is used to make a vinaigrette and is flavoured by adding strawberries, blackberries, mint, or blueberries at the beginning of fermentation.

Kiwifruit

A by-product of commercial kiwifruit growing is a large amount of waste in the form of firstly misshapen or otherwise rejected fruit that may constitute up to 30 per cent of the crop and secondly kiwifruit pomace which is the presscake residue left after kiwifruit juice manufacture. One of the uses for this waste is the production of kiwifruit vinegar. Produced commercially in New Zealand since, at least, the early 1990s Production of kiwifruit vinegar began for domestic sale in China in 2008.

Sinamak

A variation of cane vinegar from the Philippines (*sukang maasim*) is called *sinamak* which is simply a spiced version that mixes the cane vinegar with siling labuyo, onions and garlic.

Distilled Vinegar

Any type of vinegar may be distilled to produce a colourless solution of about 5% to 8% acetic acid in water. This is variously known as distilled spirit or "virgin vinegar", or white vinegar, and is used for medicinal, laboratory and cleaning purposes as well as in cooking, baking, meat preserving, including pickling. The most common starting material, because of its low cost, is malt vinegar. By FDA regulations since the 1950s, Distilled Vinegar may be synthetic ethanol made by direct chemical oxidation of wood or fossil fuels. Distilled and other highly processed vinegars are mineral deficient and, when consumed, pull calcium and other minerals from the bones and tissues.

Spirit Vinegar

The term 'spirit vinegar' is sometimes reserved for the stronger variety (5% to 20% acetic acid) made from sugar cane or from chemically produced acetic acid.

Culinary Uses

Vinegar is commonly used in food preparation, particularly in pickling processes, vinaigrettes, and other salad dressings. It is an ingredient in sauces such as mustard, ketchup, and mayonnaise. Vinegar is sometimes used while making chutneys. It is often used as a condiment. Marinades often contain vinegar.

- Condiment for beetroot — cold, cooked beetroot is commonly eaten with vinegar

- Condiment for fish and chips — People commonly use malt vinegar (or non-brewed condiment) on chips.
- Flavouring for potato chips — many American, Canadian and British manufacturers of packaged potato chips and crisps feature a variety flavoured with vinegar and salt.
- Vinegar pie — a North American dessert made with a vinegar to one's taste and similar to chess pie.
- Pickling — any vinegar can be used to pickle foods.
- Cider vinegar and sauces — cider vinegar usually is not suitable for use in delicate sauces.
- Substitute for fresh lemon juice — cider vinegar can usually be substituted for fresh lemon juice in recipes and obtain a pleasing effect although it lacks the vitamin C.
- Saucing roast lamb — pouring cider vinegar over the meat when roasting lamb, especially when combined with honey or when sliced onions have been added to the roasting pan, produces a sauce.
- Sweetened vinegar is used in the dish of pork knuckles and ginger stew which is made among Chinese people of Cantonese backgrounds to celebrate the arrival of a new child.
- Sushi rice — Japanese use rice vinegar as an essential ingredient for sushi rice.
- Red vinegar — Sometimes used in Chinese soups
- Flavouring — used in the Southern U.S. to flavour collard greens, green beans, black-eyed peas, or cabbage to taste.
- Commonly put into mint sauce, for general palate preference.
- Vinegar—especially the coconut, cane, or palm variety—is one of the principal ingredients of Philippine.

Medical Uses

Many remedies and treatments have been ascribed to vinegar over millennia and in many different cultures, however, few have been verifiable using controlled medical trials and many that are effective to some degree have significant side effects and carry the possibility of serious health risks.

Soothing for Sunburns

White vinegar applied as a spray to tissue draped over a sunburn helps restore the lost acidic level to the skin, and gives a cooling effect.

Possible Cholesterol and Triacylglycerol Effects

A 2006 study concluded that a test group of rats fed with acetic acid (the main component of vinegar) had "significantly lower values for serum total cholesterol and triacylglycerol", among other health benefits. Rats fed vinegar or acetic acid have lower blood pressure than controls, although the effect has not been tested in humans. Reduced risk of fatal ischemic heart disease was observed among participants in a trial who ate vinegar and oil salad dressings frequently.

Blood Glucose Control and Diabetic Management

Prior to hypoglycemic agents, diabetics used vinegar teas to control their symptoms. Small amounts of vinegar (approximately 20 ml or two tablespoons of domestic vinegar) added to food, or taken along with a meal, have been shown by a number of medical trials to reduce the glycemic index of carbohydrate food for people with and without diabetes. This also has been expressed as lower glycemic index ratings in the region of 30%.

Diet Control

Multiple trials indicate that taking vinegar with food increases satiety (the feeling of fullness) and so, reduces the amount of food consumed.

Infections

Vinegar has been used to fight infections since Hippocrates, who lived between 460-377 BC, prescribed it for curing persistent coughs. As a result, vinegar is popularly believed to be effective against infections. While vinegar can be an effective antibacterial cleaning agent on hard surfaces such as washroom tiles and countertops, studies show that vinegar – whether taken internally or applied topically – is not effective against infections, lice, or warts.

Researchers at the Food Biotechnology Department, Instituto de la Grasa (CSIC) in Seville, Spain, conducted research on the antimicrobial activity of several products. Vinegar and red and white wines were among the products tested. (Note: The focus of the research was olive oil, but it confirmed other findings related to vinegar and red and white wines.) The following microorganisms were used in the study: *S. aureus*, *L. monocytogenes*, *S. Enteritidis*, *E.coli* 0157:H7, *S.sonnei* and Yersinia sp. Among the items tested, vinegar (5% acetic acid) showed the strongest bactericidal activity against all strains tested, which was attributed to its high acetic acid content. The researchers noted their study confirmed previous results. It was noted that both red and white wines exhibited bactericidal activity, in particular against Salmonella Enteritidis and Yersinia sp. S. aureus and L. monocytogenes were the least sensitive to the wines.

Other Medicinal Uses

Applying vinegar to common jellyfish stings deactivates the

nematocysts; however, placing the affected areas in hot water is a more effective treatment because the venom is deactivated by heat. The latter requires immersion in 45°C (113°F) water for at least four minutes for the pain to be reduced to less than what would be accomplished using vinegar. But vinegar should not be applied to Portuguese man o' war stings, however, since they are not actually jellyfish and vinegar can cause their nematocysts to discharge venom, making the pain worse. Vinegar is often used as a natural deodorant, mainly because of its antibacterial effect.

Contrary to myth, vinegar cannot be used as a detoxification agent to circumvent urinalysis testing for cannabis.

Potential Hazards

Esophageal injury by apple cider vinegar tablets has been reported, and because vinegar products sold for medicinal purposes are neither regulated nor standardized, they vary widely in content, pH, and other respects. Long-term heavy vinegar ingestion may also cause Hypokalemia, Hyperreninemia, and Osteoporosis.

Cervical Screening Tool

Diluted vinegar 3% to 5%, has also been tested as an effective screening tool for cervical cancer. Vinegar changes the colour of affected tissue to white, making diagnosis by inspection possible, reducing in 35% the mortality for early detection against control group.

Cleaning Uses

White vinegar is often used as a household cleaning agent. Because it is acidic, it can dissolve mineral deposits from glass, coffee makers, and other smooth surfaces. For most uses dilution with water is recommended for safety and to avoid damaging the surfaces being cleaned.

Vinegar is an excellent solvent for cleaning epoxy resin and hardener, even after the epoxy has begun to harden. Malt vinegar sprinkled onto crumpled newspaper is a traditional, and still-popular, method of cleaning grease-smeared windows and mirrors in the UK. Vinegar can be used for polishing brass or bronze.

Recently, vinegar has been marketed as a green solution for many household cleaning problems. For example, vinegar has been cited recently as an eco-friendly urine cleaner for pets and as a weed killer.

Agricultural and Horticultural Uses

Herbicide Use

Vinegar can be used as an herbicide. Acetic acid is not absorbed into root systems, the vinegar will kill top growth, but perennial plants will reshoot.

Commercial vinegar, available to consumers for household use, does not exceed 5%, and solutions above 10% need careful handling since they are corrosive and damaging to skin. Stronger solutions that are labelled for use as herbicides are available from some retailers.

Miscellaneous

When a bottle of vinegar is opened, mother of vinegar may develop. It is considered harmless and can be removed by filtering.

Vinegar eels (*Turbatrix aceti*), a form of nematode that has cells that are air-borne, may occur in some forms of vinegar unless the vinegar is kept covered. These feed on the mother of vinegar and can occur in naturally fermenting vinegar. This is the reason vinegar condiment jars have tightly-fitting stoppers. Most manufacturers filter and pasteurize their product before bottling to eliminate any potential adulteration, although they are harmless when ingested.

When vinegar is added to sodium bicarbonate (baking soda), it produces a volatile mixture which rapidly decomposes into water, carbon dioxide and sodium ethanoate, which makes the reaction fizz. It is often used to illustrate typical acid-base reactions in school science experiments.

Some countries prohibit the selling of vinegar over a certain percentage acidity. As an example, the government of Canada limits the acetic acid of vinegars to between 4.1% and 12.3%.

Posca, a Roman legionaries' basic drink was vinegar mixed with water and optional honey.

According to legend, in France during the Black Plague, four thieves were able to rob houses of plague victims without being infected themselves. When finally caught, the Judge offered to grant the men their freedom, on the condition that they revealed how they managed to stay healthy. They claimed that a medicine woman sold them a potion, made of garlic soaked in soured red wine (vinegar). Variants of the recipe, called Four Thieves Vinegar, have been passed down for hundreds of years and are a staple of New Orleans Voodoo practices.

Diluted vinegar can be used as a homemade stop bath during photographic processing.

3

Production of Lactic Acid and Lactic Acid Fermentation of Food Products

Lactic acid, also known as milk acid, is a chemical compound that plays a role in several biochemical processes. It is a hydroxy carboxylic acid. It was first isolated in 1780 by the Swedish chemist Carl Wilhelm Scheele. It is a carboxylic acid, whose chemical formula $C_3H_6O_3$ has its hydroxyl group adjacent to the carboxyl group, making it an alpha hydroxy acid (AHA). In solution, it can lose a proton from the acidic group, producing the lactate ion $CH_3CH(OH)COO^-$. It is miscible with water or ethanol, and is hygroscopic.

In animals, L-lactate is constantly produced from pyruvate via the enzyme lactate dehydrogenase (LDH) in a process of fermentation during normal metabolism and exercise. It does not increase in concentration until the rate of lactate production exceeds the rate of lactate removal, which is governed by a number of factors, including monocarboxylate transporters, concentration and isoform of LDH, and oxidative capacity of tissues.

The concentration of blood lactate is usually 1–2 mmol/L at rest, but can rise to over 20 mmol/L during intense exertion. In industry, lactic acid fermentation is performed by *Lactobacillus* bacteria, among others. These bacteria can operate in the mouth; the acid they produce is responsible for the tooth decay known as caries.

In medicine, lactate is one of the main components of lactated Ringer's solution and Hartmann's solution. These intravenous fluids consist of sodium and potassium cations, with lactate and chloride anions, in solution with distilled water in concentration so as to be isotonic compared to human blood. It is most commonly used for fluid resuscitation after blood loss due to trauma, surgery, or a burn injury.

History

Lactic acid was refined for the first time by the Swedish chemist Carl Wilhelm Scheele in 1780 from sour milk. In 1808 Jons Jacob Berzelius discovered that lactic acid also is produced in muscles during exertion. Its structure was established by Johannes Wislicenus in 1873.

In 1856 Louis Pasteur discovered *Lactobacillus* and its role in the making of lactic acid. Lactic acid started to be produced commercially by the German pharmacy Boehringer Ingelheim in 1895.

Exercise and Lactate

During power exercises such as sprinting, when the rate of demand for energy is high, lactate is produced faster than the ability of the tissues to remove it, so lactate concentration begins to rise. This is a beneficial process, since the regeneration of NAD^+ ensures that energy production is maintained and exercise can continue. The increased lactate produced can be removed in a number of ways, including:

- Oxidation to pyruvate by well-oxygenated muscle cells, which is then directly used to fuel the Krebs cycle
- Conversion to glucose via gluconeogenesis in the liver and release back into the circulation.

Contrary to popular belief, this increased concentration of lactate does not directly cause acidosis, nor is it responsible for delayed onset muscle soreness. This is because lactate itself is not capable of releasing a proton, and, second, the acidic form of lactate, lactic acid, "is not produced in muscle". Analysis of the glycolytic pathway in humans indicates that there are not enough hydrogen ions present in the glycolytic intermediates to produce lactic or any other acid.

The acidosis that is associated with increases in lactate concentration during heavy exercise arises from a separate reaction. When ATP is hydrolysed, a hydrogen ion is released. ATP-derived hydrogen ions are responsible primarily for the decrease in pH. During intense exercise, aerobic metabolism cannot produce ATP quickly enough to supply the demands of the muscle. As a result, anaerobic metabolism becomes the dominant energy-producing pathway, as it can form ATP at high rates. Due to the large amounts of ATP being produced and hydrolysed in a short period of time, the buffering systems of the tissues are overcome, causing pH to fall and creating a state of acidosis, a natural process that facilitates the easier dissociation of oxyhaemoglobin and allows easier transfer of oxygen from the blood.

The effect of lactate on acidosis has been the topic of many recent conferences in the field of exercise physiology. Robergs *et al.* have accurately chased the proton movement that occurs during glycolysis. However, in doing so, they have suggested that [H^+] is an independent variable that determines its own concentration. A recent review by Lindinger *et al.* has been written to rebut the stoichiometric approach used by Robergs *et al.* In using this stoichiometric process, Robergs *et al.* have ignored the causative factors (independent variables) of the concentration of hydrogen ions (denoted [H^+]). These factors are strong ion difference [SID], PCO_2, and weak acid buffers. Lactate is a strong anion, and causes a reduction in [SID], which causes an increase in [H^+] to maintain electroneutrality. PCO_2 also causes an increase in [H^+]. During exercise, the intramuscular lactate concentration and PCO_2 increase, causing an increase in [H^+], and, thus, a decrease in pH.

Polymer Precursor

Two molecules of lactic acid can be dehydrated to lactide, a cyclic lactone. A variety of catalysts can polymerize lactide to either heterotactic or syndiotactic polylactide, which as biodegradable polyesters with valuable (*inter alia*) medical properties are currently attracting much attention. Lactic acid is used also as a monomer for producing polylactic acid (PLA), which later has developed application as biodegradable plastic. This kind of plastic is a good option for substituting conventional plastic produced from petroleum oil because of low emission of carbon dioxide. The commonly used process in producing lactic acid is via fermentation, and, later, to obtain the polylactic acid, the polymerization process follows.

Foods

Lactic acid is found primarily in sour milk products, such as koumiss, leban, yogurt, kefir, and some cottage cheeses. The casein in fermented milk is coagulated (curdled) by lactic acid. Lactic acid is also responsible for the sour flavour of sourdough breads. This acid is used in beer brewing to lower the pH and increase the body of the beer. In winemaking, a bacterial process, natural or controlled, is often used to convert the naturally present malic acid to lactic acid, to reduce the sharpness and for other flavour-related reasons. This malolactic fermentation is undertaken by the family of lactic acid bacteria.

Detergents

Lactic acid has gained importance in the detergents industry the last decade. It is a good descaler, soap-scum remover, and a registered anti-bacterial agent. It is also economically beneficial as well as part of a trend toward environmentally safer and natural ingredients.

Production of Amino Acids

Amino acids play central roles both as building blocks of proteins and as intermediates in metabolism. The 20 amino acids that are found within proteins convey a vast array of chemical versatility. The precise amino acid content, and the sequence of those amino acids, of a specific protein, is determined by the sequence of the bases in the gene that encodes that protein. The chemical properties of the amino acids of proteins determine the biological activity of the protein. Proteins not only catalyse all (or most) of the reactions in living cells, they control virtually all cellular process. In addition, proteins contain within their amino acid sequences the necessary information to determine how that protein will fold into a three dimensional structure, and the stability of the resulting structure. The field of protein folding and stability has been a critically important area of research for years, and remains today one of the great unsolved mysteries. It is, however, being actively investigated, and progress is being made every day. As we learn about amino acids, it is important to keep in mind that one of the more important reasons to understand amino acid structure and properties is to be able to understand protein structure and properties. We will see that the vastly complex characteristics of even a small, relatively simple, protein are a composite of the properties of the amino acids which comprise the protein.

Essential Amino Acids

Humans can produce 10 of the 20 amino acids. The others must be supplied in the food. Failure to obtain enough of even 1 of the 10 essential amino acids, those that we cannot make, results in degradation of the body's proteins—muscle and so forth—to obtain the one amino acid that is needed. Unlike fat and starch, the human body does not store excess amino acids for later use—the amino acids must be in the food every day.

The 10 amino acids that we can produce are alanine, asparagine, aspartic acid, cysteine, glutamic acid, glutamine, glycine, proline, serine and tyrosine. Tyrosine is produced from phenylalanine, so if the diet is deficient in phenylalanine, tyrosine will be required as well. The essential amino acids are arginine (required for the young, but not for adults), histidine, isoleucine, leucine, lysine, methionine, phenylalanine, threonine, tryptophan, and valine. These amino acids are required in the diet. Plants, of course, must be able to make all the amino acids. Humans, on the other hand, do not have all the enzymes required for the biosynthesis of all of the amino acids.

Why Learn these Structures and Properties?

It is critical that all students of the life sciences know well the structure and chemistry of the amino acids and other building blocks of biological molecules. Otherwise, it is impossible to think or talk sensibly about proteins and enzymes, or the nucleic acids.

Amino Acid Production

Background

The amino acid business is a multi-billion dollar enterprise. All twenty amino acids are sold, albeit each in greatly different quantities. Amino acids are used as animal feed additives (lysine, methionine, threonine), flavour enhancers (monosodium glutamic, serine, aspartic acid) and as speciality nutrients in the medical field. Glutamic acid, lysine and methionine account for the majority, by weight, of amino acids sold. Glutamic acid and lysine are made by fermentation; methionine is made by chemical synthesis. The major producers of amino acids are based in Japan, the US, South Korea, China and Europe.

Many microbe-based industries have their origins in traditions that go back hundreds or thousands of years. The amino acid industry has its roots in food preparation practices in Japan. Seaweeds had been used for centuries there and in other Asian countries as a flavouring ingredient. In 1908, Kikunae Ikeda of Tokyo Imperial University isolated the flavour enhancing principle from the seaweed *konbu* (also spelled *kombu, Laminaria japonica;* related to kelp) as crystals of monosodium glutamate (MSG).

Adding MSG to meat, vegetables and just about any other type of prepared food makes it savory, a property referred to as *umami*. Soon after Ikeda's discovery, and recognizing the market potential of MSG, Ajinomoto Co. in Japan began extracting MSG from acid-hydrolyzed wheat gluten or defatted soybean and selling it as a flavour enhancer.

The production of MSG via "fermentation" grew out of the ashes of WWII in Japan. Around 1957, Japanese researchers led by S. Kinoshita at Kyowa Hakko Kogyo Co. isolated soil bacteria that produced large amounts of glutamic acid. Producing strains were found by inoculating soil isolates in a grid pattern on duplicate Petri plates. The colonies were allowed to grow and one set of duplicates was killed with UV irradiation. The killed plate was overlaid with soft agar containing a *Leuconostoc mesenteroides*. Since *L. mesenteroides* required glutamic acid for growth, it only grew in the vicinity of colonies that had excreted glutamate. Potential glutamate producers were then picked from the duplicate, unkilled, plate.

Members of the *Actinobacteria* in the genus *Corynebacterium* (originally named *Micrococcus glutamicus*) were the most effective producers. Over the years, various glutamate-producing bacteria have been isolated and classified as *Arthrobacter, Brevibacterium,* or as members of other genera, but recent work has shown that almost all of these strains belong to the genus *Corynebacterium*. Wild-type cultures produced up to 10 g/l glutamic acid. Yields were quickly improved by process engineering and by developing over-producing mutants. Yields are now in excess of 100 g per liter.

The isolation of bacterial glutamate-producers led to the development of large-scale manufacture of MSG from cheap sugar and ammonia rather than from more expensive extracts of plants or animals. In the early 1960s, workers at the same company found that *C. glutamicum* homoserine auxotrophs produced lysine thus providing the first viable fermentation process for lysine production. Today these bacteria produce well over 1,000,000 metric tons of MSG and 600,000 metric tons of lysine annually.

Three general approaches are used today for making amino acids: direct chemical synthesis, fermentation and bioconversion using enzymes. Choosing between processes depends on available technology, costs of raw material, market prices and sizes, cost of running fermentation versus synthesis reactions, and the environmental impact of the process itself. The processes have not changed very much since then, but the amounts made have increased at a rate of about 2-5% per year.

Amino acids are consumed in a variety of markets. The largest by volume is the food flavouring industry. MSG, alanine, aspartate, arginine are all used to improve the flavour of food. The basis of our ability to enjoy the taste of amino acids is rooted in evolution. Animals require certain amino acids for growth and nutrition. Specific taste receptors on the human tongue have been found that are G-protein coupled receptors responsive to many of the 20 L- but not D-amino acids. For human receptors, the strongest response is for glutamate, thus providing a biochemical link to what cooks have known for centuries. Interestingly, inosine monophosphate (IMP), another well-known flavour enhancing nucleotide also sold commercially, greatly increases the response of the receptor to amino acids.

The second largest consumer of amino acids is the animal feed industry.

Lysine, methionine, threonine, tryptophan and others improve the nutritional quality of animal feeds by supplying essential amino acids that may be in low abundance in grain. Using 0.5% lysine in animal feed improves the quality of the feed as much as adding 20% soy meal. In addition, by limiting the added amino acid supplements to those required by the animal, some of the excess ammonia made via deamination reactions that is normally excreted to the environment is eliminated. The addition of microbially-produced amino acids has also increased with the onset of "Mad Cow" syndrome (BSE), a disease traced to feed supplements containing animal protein.

Finally, the pharmaceutical industry uses a variety of amino acids for making intravenous nutrient solutions for pre- and post-operative care. These mixtures account for a small percentage of the total volume of amino acids sold each year, but the requirement that they be highly purified provides a value-added component. In addition to these uses, amino acids are used as speciality chemicals in laboratories, in the manufacture of artificial sweeteners (aspartame) and in many other situations.

MSG Glutamate is the most abundant free amino acid in bacterial cytoplasm. Nevertheless, in order to be useful, glutamate producers must do two things well: they must overproduce glutamate in excess of their normal metabolic needs, and they must excrete it into culture broth. The precise mechanism by which *C. glutamicum* does these things is still not completely understood despite over forty years of study. Some physiological traits, however, are clearly involved. These include biotin auxotrophy of producing strains, a marked decrease in α- ketoglutarate dehydrogenase activity during production, and a predilection for exporting glutamate, perhaps via a specific transporter.

Many of the original glutamate-excreting strains were biotin auxotrophs, and growing in biotin deficient medium was found to "trigger" glutamate production. Biotin is a cofactor (a "vitamin") used by enzymes that carboxylate substrates. One such enzyme is acetyl-CoA carboxylase that converts Acetyl-CoA + CO_2 to Malonyl-CoA in the first step of fatty acid biosynthesis. Biotin auxotrophs growing in biotin deficient medium were proposed to have altered membranes due to suboptimal fatty acid biosynthesis. Supporting the notion of altered permeability is the observation that growth at higher temperatures, or including detergents or cell-wall biosynthesis inhibitors like penicillin in the growth medium can also trigger excretion.

Reduced levels of α-ketoglurate dehydrogenase during production may also be linked to membrane integrity.

In corynebacteria, the enzyme has three activities on two peptides: the α-kg dehydrogenase + dihydrolipoamide S-succinyltransferase peptide and the dihydrolipoamide dehydrogenase peptide. The latter is shared with pyruvate dehydrogenase and is likely to be membrane bound and thus prone to being affected by trigger factors that alter membrane composition.

Since α-ketoglurate dehydrogenase catalyses a step in the TCA cycle, the cycle is largely incomplete during glutamate production, a circumstance requiring that pools of oxaloacetate be filled, as carbon is lost through glutamate. Anaplerotic enzymes replace OAA; these enzymes include pyruvate carboxylase (Rxn. 1), malic enzyme (Rxn. 2), PEP carboxylase (Rxn. 3) and glyoxylate pathway enzymes.

1. pyruvate + CO_2 + ATP ? ? oxaloacetate + ADP + Pi
2. pyruvate + CO_2 + NADPH ? ? malate + NADP+ + H+
3. phosphoenol pyruvate + CO_2 ? ? oxaloacetate + Pi.

The actual contribution of each anaplerotic enzyme or pathway depends on the growth conditions, production phase and a host of interacting metabolic signals. However, mutants lacking pyruvate carboxylase produce as much glutamate as wild type suggesting that PEP carboxylase is the major anaplerotic enzyme whan growing on sugars. When growing on acetate or fatty acids, the glyoxylate pathway assumes a major role in filling OAA pools, although acetate is not normally used in the industrial process.

For many years, increased membrane permeability was thought to promote glutamate excretion. Recently, however, export via a specific efflux transporter has been proposed as a mechanism for avoiding exceptionally high levels of intracellular glutamate. Such an exporter has been proposed by analogy with the LysE transporter that is responsible for lysine and arginine export in *C. glutamicum*.

Other factors contributing to glutamate overproduction include metabolic flux alterations based on cell growth limitations. The following chain of events seems likely: A triggering mechanism (biotin depletion, high temperature, cell membrane alterations by detergents, oleic acid or antibiotics) results in a decrease or repression of α-kg dehydrogenase. The result is a redistribution of metabolites at the branch point in the TCA cycle leading from α-kg to succinyl-CoA or glutamate. The increase in glutamate levels, beyond that needed for cell growth, stimulates glutamate efflux as the cell attempts to maintain the proper level of intracellular glutamate. The result is the excretion of glutamate from the cell.

Industrial production of MSG Although the details of MSG production remain proprietary and differ in detail from company to company, a general outline of the industrial process is known. The process is most commonly run as a fed-batch type, where sugar is added during the fermentation process. The reason for using fed-batch rather than batch, where all components are present at the beginning of the process, is that high sugar concentrations, more that 20% in total would otherwise need to be added to the medium. Such high levels can lead to incomplete oxidation of the sugar to lactic or acetic acids or to osmotic inhibition of growth with subsequent decrease in yields.

Small seed cultures are grown on glucose, potassium phosphate, magnesium sulfate, yeast extract and urea as a source of nitrogen. Larger cultures use cheaper sources of sugar including cane or beet molasses and starch hydrolyzates from corn or cassava. The sugar source roughly parallels the geographic location of the process.

That is, corn syrup is used in the United States, tapioca (from cassava) in South Asia and cane and beet molasses in Europe and S. America. Ammonia and ammonium sulfate are used as nitrogen sources. Ammonia can also be used to control pH during fermentation. Cheap sources of vitamins and other nutrients include corn steep liquor, a by-product of cornstarch manufacture that is replete with amino acids, nucleic acids, minerals and vitamins.

To begin the process, stocks of *C. glutamicum* are used to inoculate shake flask cultures. The resulting cells are transferred to a small seed culture tank that is grown and used to inoculate a larger tank, and so on. The intermediate seed culture volumes are variable but generally in the range of 200-1000 liter, then 10,000-20,000 l and finally the production tank of about 50,000-500,000 l. The process is carefully controlled at each step such that cell density, nutrient composition, temperature, pH, aeration, agitation rates and sugar flow rates are as consistent as possible from batch to batch. Oleic acid (0.65 ml/l) might be added at the beginning of the fermentation to encourage glutamate excretion; pH is set around 8.5 with ammonia, and maintained at 7.8 during the process.

After 14 hours of growth, the temperature is increased from about 32-33°C to 35° Sugar is fed in as the fermentation proceeds up to about 36 h. During the course of the fed-batch process about 160 g or more of glucose per liter, or its equivalent, is fed into the bioreactor. Glutamate concentrations are monitored at intervals and, after the process is complete, the broth is pumped from the bioreactor to recovery tanks. Yields of glutamate from a large-scale fermentation are in excess of 100 g/l. That means that a 100,000 liter bioreactor yields about 10,000 kg of glutamate. At a market price of about $1.25 per kg, a single fermenter has a value of about $12,500. Considering the costs of energy, personnel, overhead and processing, that is not much money and is a consequence of fierce competition between companies.

L-lysine

Soon after *C. glutamicum* was commercialized for MSG production, auxotrophs were developed that excreted commercial quantities of lysine. By 1958, Kinoshita and colleagues had shown that lysine was excreted by homoserine auxotrophs. For example, a homoserine auxotroph (strain ATCC13287) was patented in 1961 that yielded 44 g lysine per liter with a conversion efficiency of 26% from sugar (g lysine/g sugar). Further increases in yield were made incrementally with the introduction of additional amino acid and vitamin auxotrophies plus development of strains resistant to antimetabolites. Typical strains today provide conversion efficiencies of over 50%.

To understand why lysine is overproduced from a physiological standpoint, one must first understand the regulatory circuits that govern lysine biosynthesis. The approaches used for creating amino acid overproducers are paradigms of rational strain improvement as contrasted with approaches that rely on random mutagenesis and screening. The pathway leading to lysine (also threonine, isoleucine, methione) biosynthesis is initiated with the conversion of aspartate to aspartyl-P via the enzyme aspartokinase (AK). The phosphorylated aspartate is then converted to aspartyl-semialdehyde (ASA) that can converted to homoserine by homoserine dehydrogenase (HSD) or to diaminopimelic acid (DAP) by a series of five enzymatic conversions, and thence to lysine. Homoserine is the starting point for making threonine and isoleucine as well as methionine. The major control points for the metabolic flux to individual amino acids occurs at the level of aspartokinase and homoserine dehydrogenase.

AK is feedback inhibited by the accumulation of threonine plus lysine. Homoserine dehydrogenase is feedback inhibited by threonine and its synthesis is repressed by methionine. Mutations that lead to the overproduction of lysine include HSD-, HSDleaky, and lysine analogue resistant mutants. HSD-mutants are homoserine auxotrophs that effectively lack the ability to make threonine, methionine and isoleucine, thus eliminating the feedback inhibition of AK by threonine. HSDleaky mutants make an HSD that is less effective at making homoserine so that threonine does not accumulate to levels sufficient to shut down AK. The benefit of the latter mutants is that additional amino acids do not need to be added to growth media to make up for their complete lack of synthesis in HSD null mutants.

Analogue resistant mutants resist the feedback inhibition effects of allosteric effectors. In the case of AK, mutants were derived that resisted analogues of lysine and/or threonine. To derive such mutants, mutagenized cultures are incubated in the presence of the analogue; those that grow "resist" the analogue and are screened for their ability to make lysine. The rationale behind the approach is that wild-type AK recognizes the lysine analog 2-aminoethyl-L-cysteine (AEC) as lysine. The enzyme is inhibited as if lysine had accumulated in the cell and the cell fails to grow due to lysine starvation. Analogue resistant mutants have AKs that do not recognize AEC as lysine; these enzymes remain uninhibited, continue to make aspartyl-P and then lysine and the cells continue to grow. Such mutants, since they no longer respond to intracellular lysine levels, continue to synthesize lysine at high levels. Other types of auxotrophs, for example for leucine and alanine also overproduce lysine since their biosynthetic pathways compete for pyruvate.

In 1996, the gene encoding a membrane protein involved in the active export of lysine– *lysE* – was cloned and sequenced. It is responsible for exporting lysine up a concentration gradient eventually leading to the accumulation of as much as 1 mole per liter of lysine. Mutations in this gene cause lysine to accumulate in the cytoplasm but not in the extracellular medium. Introducing the gene into a multi-copy plasmid in *C. glutamicum* leads to increases in the rate of efflux of intracellular lysine.

Lysine Fermentation The industrial production of L-lysine is carried out in a manner similar to that for L-glutamate production.

Sugar cane or beet molasses is used as the substrate for production. If the biotin level is low in the starting material it must be added to a level of about 30 ug/l. If an auxotroph for any amino acid is used, that amino acid must be present so that the cells may grow but should not be present at high enough concentration to exert inhibitory physiological effects such as feedback inhibition of the pathways used for production.

L-threonine

L-Threonine production has been accomplished by genetically manipulating strains of *E. coli* and stands as a good example of rational design of industrial organisms. The initial strains constructed for his purpose were derived 1969 as mutants resistant to the L-threonine analogue, α-amino-β-hydroxyvaleric acid (AHV) by Shiio & Nakamori in Japan. Since that time several strains have emerged as production strains based on a combinations of auxotrophic and analogue resistant mutations that have the collective effect of funneling carbon to L-threonine production.

Detailed understanding of threonine metabolism in *E. coli* was essential for rationally designing threonine overproducers. The regulation of threonine biosynthesis in *E. coli* is more complex than that in *C. glutamicum.*

Unlike *Corynebacterium, E. coli* has three aspartate kinases, AKI, AKII and AKIII. Two (AKI and AKII) are multidomain proteins that also have homoserine dehydrogenase activity responsible for the third step of the pathway. AKI is feedback inhibited by threonine and its synthesis is repressed by a combination of threonine and isoleucine. The synthesis of AKII is repressed by methionine. AKIII is feedback inhibited and repressed by lysine.

The second step of the pathway is catalysed by aspartate semialdehyde dehydrogenase (ASD). Its expression is repressed by an accumulation of high levels of lysine, threonine and methionine, with lysine being the most effective. The last two enzymes, homoserine kinase (HK; *thrB*) and threonine synthase (TS; *thrC*) are coexpressed along with AKI (*thrA*) as part of the *thrABC* operon. This operon is controlled by transcriptional attenuation. A 21 amino acid long leader peptide between the promoter and first gene has seven theronines and four isoleucines. The rate of transcription dictates whether the genes are transcribed. If the rate of transcription is fast (lots of charged trp-tRNA and iletRNA) then transcription is terminated; if it is slow, leading to pausing of the ribosome during translation, transcription continues and the enzymes are made.

Aside from these regulatory mechanisms, other systems also affect threonine production. These include the breakdown of threonine by threonine deaminase (*ilvA*) or threonine dehydrogenase (*tdh*), and transport systems that control the uptake or efflux of threonine from the medium.

With such a complex regulatory scheme, several approaches have been taken to "open up" the pathway to threonine production. These generally involve deriving auxotrophs for amino acids that might compete for carbon flow through the pathway, or analogue resistant mutants that have enzymes that have lost the ability to be feedback inhibited by the product of the pathway.

Analogue resistant mutants of *E coli* were derived by 1969 using the threonine analogue α-amino-β-hyroxyvaleric acid (AHV). This compound has been used for virtually all subsequent work. In addition to analogue resistance that removed the feedback inhibition on AKI, auxotrophs were made that required isoleucine. By placing the threonine operon onto multicopy plasmids and introducing it into such a strain, strains were obtained that produced about 65 g/l of threonine with yields of 48% (gthr/g sugar).

L-aspartate & L-alanine

L-Aspartate is used in foods and pharmaceuticals. The production of aspartate from immobilized *E. coli* cells has been done since 1973. Cells immobilized in various gels including polyacrylamide or κ-carrageenan, polyurethane has been the method of choice.

Aspartic acid is made by the enzyme aspartate ammonia lyase (aspartase) that carries out the following reaction in presence of ammonium fumarate:

OOCCH=CHCOO- + NH4+ ßà-OOCCH2CH(NH3+)COOOnce immobilized, the cells are quite stable retaining aspartase activity for well over 600 days even at 37°C. The process is carried out at pH 8.5 with ammonium fumarate as the substrate.

Immobilized *Pseudomonas dacunhae* cells can convert aspartate to alanine using the pyridoxalphosphate dependent aspartate β-carboxylase. The reaction proceeds as follows:

OOCCH2CH(NH3+)COO- ß à CH3CH(NH3+)COO- + CO_2

This process was industrialized in Japan in 1982 and is used today for making L-alanine.

Aspartame Aspartame is the trade name applied to a dipeptide: L-aspartyl-L-phenylananine methyl ester. It was discovered apparently by accident in 1965 by James Schlatter, a chemist working at G.D. Searle & Co. looking for new treatments for gastric ulcers. As the story goes, he spilled some intermediate on his hand and later licked his finger to pick up a piece of paper and noticed the intense sweet flavour. Aspartame is about 180 times sweeter than sucrose and, gram for gram, has about as many calories. Thus, much less is needed to provide the equivalent sweetness of sucrose. Aspartame is sold under a variety of trade names including Nutrasweet®, originally made by Searle & Co., Equal, Spoonful and Measure).

A process for making phenylalanine was developed and patented by Genex was based on older work that had demonstrated the production of phenylalanine from cinnamic acid plus ammonium ions by the activity of phenylalanine ammonia-lyase. Although known for many years, the process had not been used due to the instability of the enzyme and its inhibition of by the substrate *t*-cinnamic acid.

Genex managed to improve the stability by carrying out the bioconversion under anaerobic, static conditions, and using strains of the yeast *Rhodotorula rubra.*

The process of phenylalanine production is described in U.S. Patent #4,584,269. An initial aerobic growth phase is used to grow the yeast cells on complex medium that includes yeast extract. After growth, cells are induced to produce the enzyme by adding a substrate such as

L-, or D, L-phenylalanine or tyrosine, casein or blood hydrolyzates.

Once the enzyme has been induced, whole suspended or immobilized cells are then incubated with slow agitation under anaerobic conditions and in the presence of t-cinnamic acid (25 g/l) plus ammonium ions.

Alternatively, cinnamic acid can be added as a fed batch to the stirring cells yielding about 42.7 g/l of phenylalanine.

Other amino acids All amino acids may be produced by fermentation. Whether they will or not depends on the costs of competing technologies such as chemical synthesis or extraction from protein sources. Bacterial strains that produce amino acid are, with some exceptions, mainly derived from *Corynebacterium* sp., *Bacillus* sp. or *E. coli.*

Strains used in production are wild-type natural overproducers, auxotrophic or regulatory mutants that have altered feedback inhibition pathways, or derepressed enzyme synthesis, and/or genetically engineered organisms that have multiple copies of genes encoding rate-limiting enzymes.

Preparation of Cheese

Cheese is a generic term for a diverse group of milk-based food products. Cheese is produced throughout the world in wide-ranging flavours, textures, and forms.

Cheese consists of proteins and fat from milk, usually the milk of cows, buffalo, goats, or sheep. It is produced by coagulation of the milk protein casein. Typically, the milk is acidified and addition of the enzyme rennet causes coagulation. The solids are separated and pressed into final form. Some cheeses have molds on the rind or throughout. Most cheeses melt at cooking temperature.

Hundreds of types of cheese are produced. Their styles, textures and flavours depend on the origin of the milk (including the animal's diet), whether they have been pasteurized, the butterfat content, the bacteria and mold, the processing, and aging. Herbs, spices, or wood smoke may be used as flavoring agents. The yellow to red colour of many cheeses is from adding annatto.

For a few cheeses, the milk is curdled by adding acids such as vinegar or lemon juice. Most cheeses are acidified to a lesser degree by bacteria, which turn milk sugars into lactic acid, then the addition of rennet completes the curdling. Vegetarian alternatives to rennet are available; most are produced by fermentation of the fungus *Mucor miehei*, but others have been extracted from various species of the *Cynara* thistle family.

Cheese is valued for its portability, long life, and high content of fat, protein, calcium, and phosphorus. Cheese is more compact and has a longer shelf life than milk. Cheesemakers near a dairy region may benefit from fresher, lower-priced milk, and lower shipping costs. The long storage life of some cheese, especially if it is encased in a protective rind, allows selling when markets are favourable.

Cheese is an ancient food whose origins predate recorded history. There is no conclusive evidence indicating where cheesemaking originated, either in Europe, Central Asia or the Middle East, but the practice had spread within Europe prior to Roman times and, according to Pliny the Elder, had become a sophisticated enterprise by the time the Roman Empire came into being.

Proposed dates for the origin of cheesemaking range from around 8000 BCE (when sheep were first domesticated) to around 3000 BCE. The first cheese may have been made by people in the Middle East or by nomadic Turkic tribes in Central Asia. Since animal skins and inflated internal organs have, since ancient times, provided storage vessels for a range of foodstuffs, it is probable that the process of cheese making was discovered accidentally by storing milk in a container made from the stomach of an animal, resulting in the milk being turned to curd and whey by the rennet from the stomach. There is a legend with variations about the discovery of cheese by an Arab trader who used this method of storing milk.

Cheesemaking may have begun independently of this by the pressing and salting of curdled milk to preserve it. Observation that the effect of making milk in an animal stomach gave more solid and better-textured curds, may have led to the deliberate addition of rennet.

The earliest archeological evidence of cheesemaking has been found in Egyptian tomb murals, dating to about 2000 BCE. The earliest cheeses were likely to have been quite sour and salty, similar in texture to rustic cottage cheese or feta, a crumbly, flavourful Greek cheese.

Cheese produced in Europe, where climates are cooler than the Middle East, required less salt for preservation. With less salt and acidity, the cheese became a suitable environment for useful microbes and molds, giving aged cheeses their respective flavours.

Ancient Greece and Rome

Ancient Greek mythology credited Aristaeus with the discovery of cheese. Homer's *Odyssey* (8th century BCE) describes the Cyclops making and storing sheep's and goats' milk cheese. From Samuel Butler's translation:

"We soon reached his cave, but he was out shepherding, so we went inside and took stock of all that we could see. His cheese-racks were loaded with cheeses, and he had more lambs and kids than his pens could hold...

When he had so done he sat down and milked his ewes and goats, all in due course, and then let each of them have her own young. He curdled half the milk and set it aside in wicker strainers."

By Roman times, cheese was an everyday food and cheesemaking a mature art. Columella's *De Re Rustica* (circa 65 CE) details a cheesemaking process involving rennet coagulation, pressing of the curd, salting, and aging. Pliny's *Natural History* (77 CE) devotes a chapter (XI, 97) to describing the diversity of cheeses enjoyed by Romans of the early Empire. He stated that the best cheeses came from the villages near Nimes, but did not keep long and had to be eaten fresh. Cheeses of the Alps and Apennines were as remarkable for their variety then as now. A Ligurian cheese was noted for being made mostly from sheep's milk, and some cheeses produced nearby were stated to weigh as much as a thousand pounds each. Goats' milk cheese was a recent taste in Rome, improved over the "medicinal taste" of Gaul's similar cheeses by smoking. Of cheeses from overseas, Pliny preferred those of Bithynia in Asia Minor.

Post-classical Europe

Rome spread a uniform set of cheesemaking techniques throughout much of Europe, and introduced cheesemaking to areas without a previous history of it. As Rome declined and long-distance trade collapsed, cheese in Europe diversified further, with various locales developing their own distinctive cheesemaking traditions and products. The British Cheese

Board claims that Britain has approximately 700 distinct local cheeses; France and Italy have perhaps 400 each. (A French proverb holds there is a different French cheese for every day of the year, and Charles de Gaulle once asked "how can you govern a country in which there are 246 kinds of cheese?") Still, the advancement of the cheese art in Europe was slow during the centuries after Rome's fall. Many cheeses today were first recorded in the late Middle Ages or after— cheeses like Cheddar around 1500 CE, Parmesan in 1597, Gouda in 1697, and Camembert in 1791.

In 1546, *The Proverbs of John Heywood* claimed "the moon is made of a greene cheese." (*Greene* may refer here not to the colour, as many now think, but to being new or unaged.) Variations on this sentiment were long repeated and NASA exploited this myth for an April Fools' Day spoof announcement in 2006.

Modern Era

Until its modern spread along with European culture, cheese was nearly unheard of in oriental cultures, in the pre-Columbian Americas, and only had limited use in sub-Mediterranean Africa, mainly being widespread and popular only in Europe and areas influenced strongly by its cultures. But with the spread, first of European imperialism, and later of Euro-American culture and food, cheese has gradually become known and increasingly popular worldwide, though still rarely considered a part of local ethnic cuisines outside Europe, the Middle East, and the Americas.

The first factory for the industrial production of cheese opened in Switzerland in 1815, but it was in the United States where large-scale production first found real success. Credit usually goes to Jesse Williams, a dairy farmer from Rome, New York, who in 1851 started making cheese in an assembly-line fashion using the milk from neighboring farms. Within decades hundreds of such dairy associations existed.

The 1980s saw the beginnings of mass-produced rennet, and by the turn of the century scientists were producing pure microbial cultures. Before then, bacteria in cheesemaking had come from the environment or from recycling an earlier batch's whey; the pure cultures meant a more standardized cheese could be produced. Factory-made cheese overtook traditional cheesemaking in the World War II era, and factories have been the source of most cheese in America and Europe ever since. Today, Americans buy more processed cheese than "real", factory-made or not.

Making Cheese

Curdling

A required step in cheesemaking is separating the milk into solid curds and liquid whey. Usually this is done by acidifying (souring) the milk and adding rennet. The acidification can be accomplished directly by the addition of an acid like vinegar in a few cases (paneer, queso fresco), but usually starter bacteria are employed instead. These starter bacteria convert milk sugars into lactic acid. The same bacteria (and the enzymes they produce) also play a large role in the eventual flavour of aged cheeses. Most cheeses are made with starter bacteria from the *Lactococci*, *Lactobacilli*, or *Streptococci* families. Swiss starter cultures also include *Propionibacter shermani*, which produces carbon dioxide gas bubbles during aging, giving Swiss cheese or Emmental its holes (called eyes").

Some fresh cheeses are curdled only by acidity, but most cheeses also use rennet. Rennet sets the cheese into a strong and rubbery gel compared to the fragile curds produced by acidic coagulation alone. It also allows curdling at a lower acidity—important because flavour-making bacteria are inhibited in high-acidity environments. In general, softer, smaller, fresher cheeses are curdled with a greater proportion of acid to rennet than harder, larger, longer-aged varieties.

Curd Processing

At this point, the cheese has set into a very moist gel. Some soft cheeses are now essentially complete: they are drained, salted, and packaged. For most of the rest, the curd is cut into small cubes. This allows water to drain from the individual pieces of curd.

Some hard cheeses are then heated to temperatures in the range of 35–55 °C (95–131 °F). This forces more whey from the cut curd. It also changes the taste of the finished cheese, affecting both the bacterial culture and the milk chemistry. Cheeses that are heated to the higher temperatures are usually made with thermophilic starter bacteria that survive this step—either *lactobacilli* or *streptococci*.

Salt has roles in cheese besides adding a salty flavour. It preserves cheese from spoiling, draws moisture from the curd, and firms cheese's texture in an interaction with its proteins. Some cheeses are salted from the outside with dry salt or brine washes. Most cheeses have the salt mixed directly into the curds.

Other techniques influence a cheese's texture and flavour. Some examples:

- Stretching: (Mozzarella, Provolone) The curd is stretched and kneaded in hot water, developing a stringy, fibrous body.
- Cheddaring: (Cheddar, other English cheeses) The cut curd is repeatedly piled up, pushing more moisture away. The curd is also mixed (or *milled*) for a long time, taking the sharp edges off the cut curd pieces and influencing the final product's texture.

- Washing: (Edam, Gouda, Colby) The curd is washed in warm water, lowering its acidity and making for a milder-tasting cheese.

Most cheeses achieve their final shape when the curds are pressed into a mold or form. The harder the cheese, the more pressure is applied. The pressure drives out moisture—the molds are designed to allow water to escape—and unifies the curds into a single solid body.

Ripening

A newborn cheese is usually salty yet bland in flavour and, for harder varieties, rubbery in texture. These qualities are sometimes enjoyed—cheese curds are eaten on their own—but normally cheeses are left to rest under controlled conditions. This aging period (also called ripening, or, from the French, *affinage*) lasts from a few days to several years. As a cheese ages, microbes and enzymes transform texture and intensify flavour. This transformation is largely a result of the breakdown of casein proteins and milkfat into a complex mix of amino acids, amines, and fatty acids.

Some cheeses have additional bacteria or molds intentionally introduced before or during aging. In traditional cheesemaking, these microbes might be already present in the aging room; they are simply allowed to settle and grow on the stored cheeses. More often today, prepared cultures are used, giving more consistent results and putting fewer constraints on the environment where the cheese ages. These cheeses include soft ripened cheeses such as Brie and Camembert, blue cheeses such as Roquefort, Stilton, Gorgonzola, and rind-washed cheeses such as Limburger.

Types

Factors in Categorization

Factors relevant to the categorization of cheeses include:

- Length of aging
- Texture
- Methods of making
- Fat content
- Kind of milk
- Country/Region of Origin.

List of Common Categories

No one categorization scheme can capture all the diversity of the world's cheeses. In practice, no single system is employed and different factors are emphasised in describing different classes of cheeses. This typical list of cheeses includes categories from food writer Barbara Ensrud.

- fresh
- whey
- pasta filata
- semi-soft
- semi-firm
- hard
- double and triple cream
- soft-ripened
- blue vein
- goat or sheep
- sharp
- processed.

Fresh, whey and Stretched Curd Cheeses

The main factor in the categorization of these cheese is their age. Fresh cheeses without additional preservatives can spoil in a matter of days. For these simplest cheeses, milk is curdled and drained, with little other processing. Examples include cottage cheese, Romanian Ca°, Neufchâtel (the model for American-style cream cheese), and fresh goat's milk chèvre. Such cheeses are soft and spreadable, with a mild taste.

Whey cheeses are fresh cheeses made from the whey discarded while producing other cheeses. Provençal Brousse, Corsican Brocciu, Italian Ricotta, Romanian Urda, Greek Mizithra, and Norwegian Geitost are examples. Brocciu is mostly eaten fresh, and is as such a major ingredient in Corsican cuisine, but it can be aged too. Traditional *pasta filata* cheeses such as Mozzarella also fall into the fresh cheese category. Fresh curds are stretched and kneaded in hot water to form a ball of Mozzarella, which in southern Italy is usually eaten within a few hours of being made. Stored in brine, it can be shipped, and is known worldwide for its use on pizzas. Other firm fresh cheeses include paneer and queso fresco.

Classed by Texture

Categorizing cheeses by firmness is a common but inexact practice. The lines between "soft", "semi-soft", "semi-hard", and "hard" are arbitrary, and many types of cheese are made in softer or firmer variations. The factor that controls cheese hardness is moisture content, which depends on the pressure with which it is packed into molds, and aging time.

Semi-soft cheeses and the sub-group, *Monastery* cheeses have a high moisture content and tend to be bland in flavour. Some well-known varieties include Havarti, Munster and Port Salut.

Cheeses that range in texture from semi-soft to firm include Swiss-style cheeses like Emmental and Gruyère. The same bacteria that give such cheeses their eyes also contribute to their aromatic and sharp flavours. Other semi-soft to firm cheeses include Gouda, Edam, Jarlsberg and Cantal. Cheeses of this type are ideal for melting and are used on toast for quick snacks.

Harder cheeses have a lower moisture content than softer cheeses. They are generally packed into molds under more pressure and aged for a longer time. Cheeses that are semi-hard to hard include the familiar Cheddar, originating in the village of Cheddar in England but now used as a generic term for this style of cheese, of which varieties are imitated worldwide and are marketed by strength or the length of time they have been aged. Cheddar is one of a family of semi-hard or hard cheeses (including Cheshire and Gloucester) whose curd is cut, gently heated, piled, and stirred before being pressed into forms. Colby and Monterey Jack are similar but milder cheeses; their curd is rinsed before it is pressed, washing away some acidity and calcium. A similar curd-washing takes place when making the Dutch cheeses Edam and Gouda.

Hard cheeses — "grating cheeses" such as Parmesan and Pecorino Romano—are quite firmly packed into large forms and aged for months or years.

Classed by Content

Some cheeses are categorized by the source of the milk used to produce them or by the added fat content of the milk from which they are produced. While most of the world's commercially available cheese is made from cows' milk, many parts of the world also produce cheese from goats and sheep, well-known examples being Roquefort, produced in France, and Pecorino Romano, produced in Italy, from ewe's milk. One farm in Sweden also produces cheese from moose's milk. Sometimes cheeses of a similar style may be available made from milk of different sources- Feta style cheeses, for example, are made from goats' milk in Greece and from sheep and cows' milk elsewhere.

Double cream cheeses are soft cheeses of cows' milk enriched with cream so that their fat content is 60% or, in the case of triple creams, 75%.

Soft-ripened and Blue-vein

There are three main categories of cheese in which the presence of mold is a significant feature: soft ripened cheeses, washed rind cheeses and blue cheeses.

Soft-ripened cheeses begin firm and rather chalky in texture, but are aged from the exterior inwards by exposing them to mold. The mold may be a velvety *bloom* of *Penicillium candida* or *P. camemberti* that forms a flexible white crust and contributes to the smooth, runny, or gooey textures and more intense flavours of these aged cheeses. Brie and Camembert, the most famous of these cheeses, are made by allowing white mold to grow on the outside of a soft cheese for a few days or weeks. Goats' milk cheeses are often treated in a similar manner, sometimes with white molds (Chèvre-Boite) and sometimes with blue.

Washed-rind cheeses are soft in character and ripen inwards like those with white molds; however, they are treated differently. Washed rind cheeses are periodically cured in a solution of saltwater brine and other mold-bearing agents that may include beer, wine, brandy, and spices, making their surfaces amenable to a class of bacteria *Brevibacterium linens* (the reddish-orange "smear bacteria") that impart pungent odors and distinctive flavours. Washed-rind cheeses can be soft (Limburger), semi-hard (Munster), or hard (Appenzeller). The same bacteria can also have some impact on cheeses that are simply ripened in humid conditions, like Camembert.

So-called blue cheese is created by inoculating a cheese with *Penicillium roqueforti* or *Penicillium glaucum*. This is done while the cheese is still in the form of loosely pressed curds, and may be further enhanced by piercing a ripening block of cheese with skewers in an atmosphere in which the mold is prevalent. The mold grows within the cheese as it ages. These cheeses have distinct blue veins, which gives them their name and, often, assertive flavours.

The molds range from pale green to dark blue, and may be accompanied by white and crusty brown molds. Their texture can be soft or firm. Some of the most renowned cheeses are of this type, each with its own distinctive colour, flavour, texture and smell. They include Roquefort, Gorgonzola, and Stilton.

Processed Cheeses

Processed cheese is made from traditional cheese and emulsifying salts, often with the addition of milk, more salt, preservatives, and food colouring. It is inexpensive, consistent, and melts smoothly. It is sold packaged and either pre-sliced or unsliced, in a number of varieties. It is also available in aerosol cans in some countries.

Eating and Cooking

At refrigerator temperatures, the fat in a piece of cheese is as hard as unsoftened butter, and its protein structure is stiff as well. Flavour and odor compounds are less easily liberated when cold. For improvements in flavour and texture, it is widely advised that cheeses be allowed to warm up to room temperature before eating. If the cheese is further warmed, to 26–32 °C (79–90 °F), the fats will begin to "sweat out" as they go beyond soft to fully liquid.

Above room temperatures, most hard cheeses melt. Rennet-curdled cheeses have a gel-like protein matrix that is broken down by heat. When enough protein bonds are broken, the cheese itself turns from a solid to a viscous liquid. Soft, high-moisture cheeses will melt at around 55 °C (131 °F), while hard, low-moisture cheeses such as Parmesan remain solid until they reach about 82 °C (180 °F). Acid-set cheeses, including halloumi, paneer, some whey cheeses and many varieties of fresh goat cheese, have a protein structure that remains intact at high temperatures. When cooked, these cheeses just get firmer as water evaporates.

Some cheeses, like raclette, melt smoothly; many tend to become stringy or suffer from a separation of their fats. Many of these can be coaxed into melting smoothly in the presence of acids or starch. Fondue, with wine providing the acidity, is a good example of a smoothly melted cheese dish. Elastic stringiness is a quality that is sometimes enjoyed, in dishes including pizza and Welsh rarebit. Even a melted cheese eventually turns solid again, after enough moisture is cooked off. The saying "you can't melt cheese twice" (meaning "some things can only be done once") refers to the fact that oils leach out during the first melting and are gone, leaving the non-meltable solids behind.

As its temperature continues to rise, cheese will brown and eventually burn. Browned, partially burned cheese has a particular distinct flavour of its own and is frequently used in cooking (e.g., sprinkling atop items before baking them).

Health and Nutrition

In general, cheese supplies a great deal of calcium, protein, phosphorus and fat. A 30-gram (1.1 oz) serving of Cheddar cheese contains about 7 grams (0.25 oz) of protein and 200 milligrams of calcium. Nutritionally, cheese is essentially concentrated milk: it takes about 200 grams (7.1 oz) of milk to provide that much protein, and 150 grams (5.3 oz) to equal the calcium.

Cheese potentially shares other nutritional properties of milk. The Centre for Science in the Public Interest describes cheese as America's number one source of saturated fat, adding that the average American ate 30 lb (14 kg) of cheese in the year 2000, up from 11 lb (5 kg) in 1970. Their recommendation is to limit full-fat cheese consumption to 2 oz (57 g) a week. Whether cheese's highly saturated fat actually leads to an increased risk of heart disease is called into question when considering France and Greece, which lead the world in cheese eating (more than 14 oz/400 g a week per person, or over 45 lb/20 kg a year) yet have relatively low rates of heart disease. This seeming discrepancy is called the French paradox; the higher rates of consumption of red wine in these countries is often invoked as at least a partial explanation.

Some studies claim that cheddar, mozzarella, Swiss and American cheeses can help to prevent tooth decay. Several mechanisms for this protection have been proposed:

- The calcium, protein, and phosphorus in cheese may act to protect tooth enamel.
- Cheese increases saliva flow, washing away acids and sugars.

Controversy

Effect on Sleep

A study by the British Cheese Board in 2005 to determine the effect of cheese upon sleep and dreaming discovered that, contrary to the idea that cheese commonly causes nightmares, the effect of cheese upon sleep was positive. The majority of the two hundred people tested over a fortnight claimed beneficial results from consuming cheeses before going to bed, the cheese promoting good sleep. Six cheeses were tested and the findings were that the dreams produced were specific to the type of cheese. Although the apparent effects were in some cases described as colourful and vivid, or cryptic, none of the cheeses tested were found to induce nightmares. However, the six cheeses were all British. The results might be entirely different if a wider range of cheeses were tested. Cheese contains tryptophan, an amino acid that has been found to relieve stress and induce sleep.

Casein

Like other dairy products, cheese contains casein, a substance that when digested by humans breaks down into several chemicals, including casomorphine, an opioid peptide. In the early 1960s it was hypothesized that autism can be caused or aggravated by opioid peptides. Studies supporting these claims have had significant flaws, so the data are inadequate to guide autism treatment recommendations.

Lactose

Cheese is often avoided by those who are lactose intolerant, but ripened cheeses like Cheddar contain only about 5% of the lactose found in whole milk, and aged cheeses contain almost none. Nevertheless, people with severe lactose intolerance should avoid eating dairy cheese. As a natural product, the same kind of cheese may contain different amounts of lactose on different occasions, causing unexpected painful reactions. As an alternative, also for vegans, there is already a wide range of different soy cheese kinds available. Some people suffer reactions to amines found in cheese, particularly histamine and tyramine. Some aged cheeses contain significant concentrations of these amines, which can trigger symptoms mimicking an allergic reaction: headaches, rashes, and blood pressure elevations.

Pasteurization

A number of food safety agencies around the world have warned of the risks of raw-milk cheeses. The U.S. Food and Drug Administration states that soft raw-milk cheeses can cause "serious infectious diseases including listeriosis, brucellosis, salmonellosis and tuberculosis". It is U.S. law since 1944 that all raw-milk cheeses (including imports since 1951) must be aged at least 60 days. Australia has a wide ban on raw-milk cheeses as well, though in recent years exceptions have been made for Swiss Gruyère, Emmental and Sbrinz, and for French Roquefort. There is a trend for cheeses to be pasteurized even when not required by law.

Compulsory pasteurization is controversial. Pasteurization does change the flavour of cheeses, and unpasteurized cheeses are often considered to have better flavour, so there are reasons not to pasteurize all cheeses. Some say that health concerns are overstated, pointing out that milk pasteurization does not ensure cheese safety.

Pregnant women may face an additional risk from cheese; the U.S. Centres for Disease Control has warned pregnant women against eating soft-ripened cheeses and blue-veined cheeses, due to the listeria risk, which can cause miscarriage or harm to the fetus during birth.

World Production and Consumption

Worldwide, cheese is a major agricultural product. According to the Food and Agricultural Organization of the United Nations, over 18 million metric tons of cheese were produced worldwide in 2004. This is about three kilos for each person on Earth. The largest producer of cheese is the United States, accounting for 30% of world production, followed by Germany and France.

The biggest exporter of cheese, by monetary value, is France; the second, Germany (although it is first by quantity). Among the top ten exporters, only Ireland, New Zealand, the Netherlands and Australia have a cheese production that is mainly export oriented: respectively 95%, 90%, 72%, and 65% of their cheese production is exported. Only 30% of French production, the world's largest exporter, is exported. The United States, the biggest world producer of cheese, is a marginal exporter, as most of its production is for the domestic market.

Germany is the largest importer of cheese. The UK and Italy are the second- and third-largest importers.

Greece is the world's largest (per capita) consumer of cheese, with 30.0 kg eaten by the average Greek. (Feta accounts for three-quarters of this consumption.) France is the second biggest consumer of cheese, with 24 kg by inhabitant. Emmental (used mainly as a cooking ingredient) and Camembert are the most common cheeses in France Italy is the third biggest consumer by person with 22.9 kg. In the U.S., the consumption of cheese is quickly increasing and has nearly tripled between 1970 and 2003. The consumption per person has reached, in 2003, 14.1 kg (31 lb). Mozzarella is America's favorite cheese and accounts for nearly a third of its consumption, mainly because it is one of the main ingredients of pizza.

Microbial Pectic Enzymes Clarification of Fruit Juice

Several means of preserving and clarifying fruit juices are known. A usual way in which they are kept from fermenting is by the use of salcyli or benzoic acid, or some other antiseptic substance, which kills the fermentative germ, or arrests its action for a long time. About two ounces of salicylic acid, previously dissolved in some alcohol, to twenty-five gallons of juice or ten grains to a quart, will be the proper quantity. Another method quite practicable is to fill the freshly prepared cold juice in bottles until it reaches into the neck of the bottle, and on the top of this fruit juice, carefully using the neck of the bottle as a guide, a little glycerine is placed. Juices thus preserved will keep in an unchanged condition in any season.

Probably the best method of preserving fruit juices is to add fifteen per cent, of alcohol of 95°. On this addition mucilaginous matter and albumen will be deposited. The juices can then be kept in demijohns or barrels, tightly closed, and when at rest, they become perfectly clear, so that clarification is unnecessary provided the vessel has not been moved. The liquid is then syphoned off with a hose.

Fruit juice may be clarified by heating it either alone or mixed with a small quantity of egg albumen, in an enameled vessel, that has a tight-fitting lid to close, without stirring, to near the boiling point of water, until the impurities have coagulated, which will occur at 70° C. (158° F.), and either risen to the top or sunk to the bottom. Then filter through felt or flannel bag.

This heating process will entirely destroy the germs of fermentation, and the subsequent filtration will clarify the juices, which should be kept in a cool place, and will remain unchanged, and keep indefinitely if properly and carefully prepared. If the corks are coated or the neck of the bottles dipped in melted paraffine to prevent access of air through the pores, it is a wise precaution. The utmost care must be observed in heating the juices; heat no more than absolutely required, as the delicate aroma would considerably suffer.

Lime And Lemon Juice

Of late years the expressed juice of the closely allied fruits, limes and lemons, which we have already described in the previous chapter, have been extensively employed by manufacturers of carbonated beverages. Great stress has been laid on the improved quality of the goods of which they form a component part. Where intelligence and experience direct their use, the harmful effects, sometimes ascribed to them, are not likely to occur. The objection most frequently urged by many against lime or lemon juice is its tendency to create a cloudy or ropy appearance in the beverage. In using lemon or lime juice, it is well applied precaution to see that it is, first- the genuine article; and, second- that it is refiltered carefully before using. Deception is practiced in most cases, as shiploads of this article are manufactured in Europe from tartaric acid, citric acid and a good addition of sulphuric and acetic acids to make it more acid. It seems that its being imported, or represented as such, gains lime juice the full confidence of the bottler, whose honest intention is to use a fine article. The fruits are sliced and then squeezed in wooden presses, the juice being run into puncheons and quickly bunged up. This is a most important point in preparing the juice in a tropical climate, for if left exposed it would rapidly decompose. None but the choicest fruit should be used, and only about two-thirds the juice pressed out, thus insuring greater freedom from mucilaginous and pulpy matter. The further pressings, together with the juice of the unsound fruit, is evaporated to the consistence of molasses and shipped away for the manufacture of citric acid. It is stated that limes and lemons may be preserved by the very simple process of varnishing them with a solution of shellac in alcohol. Fresh juice is thus obtainable at all seasons.

Lime juice is very variable as to quality, which depends upon the method of extraction and the quality of the fruit. It contains citric acid, gum, sugar, albumen, extractive matter, inorganic salts and water. The most important and valuable constituent is the citric acid. There is only a mere trace of sugar, while the quantity of gum and albumen is much less than that contained in lemon juice, on which account some claim it is much less liable to fermentation and decomposition than the latter. The quantity of inorganic salts contained in lime juice is about the same, and is also of the same nature as is obtained from lemon juice.

According to good authority a good quality of lime juice should contain not less than 7.25 per cent, of free citric acid.

Lemon juice, owing to the fact that it contains much more sugar and mucilage than lime juice, is more liable to fermentation and decomposition, and the addition of at least one-tenth part alcohol will act as a preventive. Its clarification tends to remove the fermentative germs, and in this state it should be used in compounding carbonated drinks. Lemon juice is frequently adulterated with water, sugar or gum, and sulphuric or acetic acid. The modus operandi is to dilute the genuine juice with water, and then bring up the density with the sugar, or gum, and the percentage of acid with one or the other of the above acids. To guard against the catastrophies invariably following the use of adulterated or inferior grade goods, it is always advisable to purchase of none but reputable houses, whose standing as business men and manufacturers is above reproach. Both lime and lemon juice are so often spurious that this conservative course should be pursued, no matter what the price or inducement offered. Notwithstanding the prejudice that some carbohators bear against lime or lemon juice, the best quality of drinks now contain them. Citric and tartaric acids are good in their way; but, according to the testimony of experienced bottlers, there is a certain indefinable richness given beverages by the former articles which the latter cannot approach.

Preservation of Lime and Lemon Juice

This is a matter of great importance, which has been settled according to the individual preferences of the manufacturer. Salicylic acid, sodium salicylate and bisulphite of lime, are sometimes used as preservatives for lime and lemon juice, also alcohol is employed to fortify the juice. The finest quality of juice may keep for some time without any preservative, it being filled into clean dry bottles after having been allowed to deposit, and securely corked. If evaporated to dryness, the residue is apt to become musty. Sugar is undesirable, on account of increase of bulk. It is obviously often very desirable to have lime-juice in a highly concentrated yet instantly available state. If to 100 ounces of lime or lemon juice, fifteen ounces of glycerine be added, and the whole evaporated in vacuo to twenty-five ounces, we get a juice concentrated to one-fourth, with perfect keeping power. If alcohol be added to the concentrated juice, much of the vegetable matter is thrown out of solution, which may probably deteriorate its power; with glycerine this is not the case, the resulting product being a most stable and elegant one. The best way to preserve non-concentrated lime or lemon juice is to add ten per cent, of alcohol of 95° to the fresh juice, and to bottle and seal.

Bacillus firmus –I-4071 produced very high level of polygalacturonase by solid state fermentation (SSF). Fifty one bacterial isolates were isolated from fermented clayed Solanum tuberosum peels collected from different restaurants in Kalubeia governorate, Egypt. All bacterial isolates were screened for their ability to produce pectinases using apple pectin under solid state fermentation (SSF) conditions. The results showed that all of these isolates were found to have appreciable pectinolytic productivities of which twenty isolates showed good pectinases-producing potentialities using agro-industrial wastes viz. Solanum tuberosum (ST), Solanum melanogena (SM), Echornia crasips (EC) and citrus peels mixture (CPM) at 30°C and pH 6 by pectin clearing zone (PCZ) technique. Three bacterial isolates, viz: 4071, 107 and 10104 were found to exhibit a higher polygalacturonase (PG) production by attacking Solanum tuberosum (ST) peels compared to other wastes. The three most potent bacterial isolates were identified on the bases of cell shape, cell arrangement, relation to oxygen and physiological and biochemical tests as Bacillus firmus, I-4071, B. firmus-I-10104 and Bacillus laterosporus-I-107.

The optimum inoculum size for production of polygalacturonase production by B. firmus-I-10104 on Solanum tuberosum (ST) peels was 1 ml (30 x 10 CFU); substrate concentration, 1.25 g/25 ml; incubation period, 15-96 hours, pH, 6.0; incubation temperature, 37 °C; different nitrogen source; peptone (0.1 g/l); different carbon source, control; different amine acids, control and finally without any vitamins. These results suggesting that, the polygalacturonase was produced from cheap raw material under solid state fermentation under all optimal conditions for application in the clarification of juice.

Fungal and Bacterial Cellulases

Cellulase refers to a class of enzymes produced chiefly by fungi, bacteria, and protozoans that catalyse cellulolysis (i.e. the hydrolysis) of cellulose). However, there are also cellulases produced by other types of organisms such as plants and animals. Several different kinds of cellulases are known, which differ structurally and mechanistically. The EC number for this group of enzymes is EC 3.2.1.4.

Reaction: Hydrolysis of 1,4-beta-D-glycosidic linkages in cellulose, lichenin and cereal beta-D-glucans.

Other names 'endoglucanases' are: endo-1,4-beta-glucanase, carboxymethyl cellulase (CMCase), endo-1,4-beta-D-glucanase, Beta-1,4-glucanase, Beta-1,4-endoglucan hydrolase, Celludextrinase. The other types of cellulase belong to Excocellulases. Bet glucosidases can also be considered as yet another group of cellulases. The expression 'Avicelase' refers to the total cellulase activity of a given sample of the enzyme(s). The cellulase may be the result of the action of more than one type of enzymes.

Types and Action

Five general types of cellulases based on the type of reaction catalysed:

- Endo-cellulase breaks internal bonds to disrupt the crystalline structure of cellulose and expose individual cellulose polysaccharide chains
- Exo-cellulase cleaves 2-4 units from the ends of the exposed chains

produced by endocellulase, resulting in the tetrasaccharides or disaccharide such as cellobiose. There are two main types of exo-cellulases (or cellobiohydrolases, abbreviate CBH)- one type working processively from the reducing end, and one type working processively from the non-reducing end of cellulose.

- Cellobiase or beta-glucosidase hydrolyses the exo-cellulase product into individual monosaccharides.
- Oxidative cellulases that depolymerize cellulose by radical reactions, as for instance cellobiose dehydrogenase (acceptor).
- Cellulose phosphorylases that depolymerize cellulose using phosphates instead of water.

In the most familiar case of cellulase activity, the enzyme complex breaks down cellulose to beta-glucose. This type of cellulase is produced mainly by symbiotic bacteria in the ruminating chambers of herbivores. Aside from ruminants, most animals (including humans) do not produce cellulase in their bodies and can only partially break down cellulose through fermentation, limiting their ability to use energy in fibrous plant material.

Enzymes that hydrolyze hemicellulose are usually referred to as hemicellulase and are usually classified under cellulase in general. Enzymes that cleave lignin are occasionally classified as cellulase, but this is usually considered erroneous. Within the above types there are also progressive (also known as processive) and non-progressive types. Progressive cellulase will continue to interact with a single polysaccharide strand, non-progressive cellulase will interact once then disengage and engage another polysaccharide strand.

Most fungal cellulases have a two-domain structure with one catalytic domain, and one cellulose binding domain, that are connected by a flexible linker. This structure is adaption for working on an insoluble substrate and it allows the enzyme to diffuse two-dimensionally on a surface in a caterpillar way. However, there are also cellulases (mostly endoglucanases) that lacks cellulose binding domain. These enzymes might have a swelling function.

Cellulase Production

Successful utilization of cellulosic materials as renewable carbon sources is dependent on the development of economically feasible process technologies for cellulase production, and for the enzymatic hydrolysis of cellulosic materials to low molecular weight products such as hexoses and pentoses.

Spano et al. (4) showed that cellulase production, was the most expensive step during ethanol production from cellulosic biomass, in that it accounted for approximately 40% of the total cost. Significant cost reduction is required in order to enhance the commercial viability of cellulase production technology.

A cellulosic enzyme system consists of three major components: endo-ß-glucanase (EC 3.2.1.4), exo-ß-glucanase (EC 3.2.1.91) and ß-glucosidase (EC 3.2.1.21). The mode of action of each of these being:

(1) Endo-p-glucanase, 1,4-ß-D-glucan glucanohydrolase, CMCase, Cx: "random" scission of cellulose chains yielding glucose and cello-oligo saccharides.

(2) Exo-P-glucanase, 1,4-ß - D-glucan cellobiohydrolase, Avicelase, C1: exo-attack on the non-reducing end of cellulase with cellobiose as the primary structure.

(3) ß-glucosidase, cellobiase: hydrolysis of cellobiose to glucose.

Reese et al. 1950 (5) proposed that exo-ß-glucanase causes a disruption in cellulose hydrogen bonding, followed by hydrolysis of the accessible cellulose with endo-ß-gucanase. According to this hypothesis, in a synergistic sequence of events, endo-ß-glucanase acts randomly on the cellulose chain, while exo-ß-glucanase acts on exposed chain ends by splitting off cellobiose or glucose. Cellobiose is subsequently hydrolysed by p-glucosidase to glucose. This hypothesis is however the opposite of that proposed by Reese et al. (5), and indicates that three, rather that two enzymes are essential for the decomposition of cellulosic biomass.

Screening of Cellulase-producing Microorganisms

Although a large number of microorganisms are capable of degrading cellulose, only a few of these microorganisms produce significant quantities of cell-free enzymes capable of completely hydrolysing crystalline cellulose *in vitro*. Fungi are the main cellulase-producing microorganisms, though a few bacteria and actinomycetes have also been recently reported to yield cellulase activity. Microorganisms of the genera *Trichoderma* and *Aspergillus* are thought to be cellulase producers, and crude enzymes produced by these microorganisms are commercially available for agricultural use. Microorganisms of the genus *Trichoderma* produce relatively large quantities of endo-ß-glucanase and exo-ß-glucanase, but only low levels of ß-glucosidase, while those of the genus *Aspergillus* produce relatively large quantities of endo-ß-glucanase and ß-glucosidase with low levels of exo-ß- glucanase production.

In order to obtain a high titer cellulase-producing strain, various cellulase-producing strains obtained from Japanese and overseas research institutions were screened in our laboratory. On the basis of this screening, *Trichoderma reesei* QM-9414, isolated by the U.S. Army Natick Laboratory, was selected as the best cellulase-producing strain and was subjected to mono-colony isolation in order to obtain strain KY-746, which was used as a parent strain. A plate screening system which enables the isolation and evaluation of mutants was initially investigated, which led to the development of methodology for producing high titer cellulase.

Fungal strains of the genus *Trichoderma,* when cultured on an agar medium, are generally difficult to isolate, owing to the lateral spreading of hyphae over the surface. Hyphal growth was controlled by the addition of Triton X-100[7] and L-sorbose, an inhibitor of P-l,3-glucan biosynthesis, to the medium, thus enabling a large number of strains to be screened.. Large numbers of mutant strains exhibiting a diversity of characteristics were obtained by UV and nitrosoguanidine (NTG) treatment and various other techniques. A strain free from catabolite repression for example, was induced using a culture medium containing glycerol and glucose, while a high titer strain was induced by addition of L-sorbose to the medium. In addition, a strain of elevated (3-glucosidase activity was isolated from medium containing 4-methyl-umbelliferyl-ß-D-glucoside.

Strain Improvement for Cellulase Production

The commercial use of cellulases is dependent on the following:

i) high titer and good enzymatic activity

ii) low production cost, and

iii) feasible mass production.

Development of a Process for High-titer Cellulase Production

Representative mutants derived from KY-746 were batch cultured in a 5-L fermentor. Considerable enhancement of titer was observed in four strains: KDG-3, PC-3-7, PCD-10, and CDU-11. An evaluation of carbon source utilization by these microorganisms, suggested that the production of high-titer cellulases necessitated an increase in the concentration of Avicel, the carbon source.

A somewhat lower than expected maximum cellulase titer was however obtained at Avicel concentrations greater than 6% in batch culture. This relatively lower titer, may be attributed either to adsorption of the cellulase produced on to the Avicel, or rate limitation of aeration due to the high viscosity of the culture medium.

In order to overcome this, semi-batch culturing was investigated, where it was found that an initial Avicel concentration of 6%, followed by subsequent addition of 4% Avicel together with other components after a few days, resulted in relatively higher titer cellulase than that obtained in the batch process. Titer improvement by the semi-batch process was however dependent on the types of microorganisms used. While the titer of KY-746 was relatively unaffected by the semi-batch process, various degrees of improvement in titer were observed for KDG-3, PC-3-7, PCD-10, and CDU-11. This semi-batch process was further utilized in order to optimize the stirring rate, aeration volume, feed change, seed volume, and the seed culture period.

T. reesei generally exhibits poor (3-glucosidase activity. Strain CDU-11 was therefore created in order to enhance p-glucosidase activity. P-glucosidase activity can also be induced by elaborately controlling culture conditions.

CMCase, Avicelase, and FPU activities exhibited a pH optimum of approximately 4, while the pH optimum of P-glucosidase was between pH 5 and 6. pH adjustment was therefore investigated with the objective of uniformly inducing all three activities.

Culturing at pH 4 for 2 days, followed by a shift to either pH 5 or 6 on the third day, resulted in a marked increase in the p-glucosidase activity of PC-3-7, but no beneficial effect was observed in the case of CDU-11 and of KDG-3, which was already of high [3-glucosidase activity.

On the basis of these findings, a scale-up experiment was conducted in a 1-kL tank using the semi-batch culturing method with 10% Avicel as a carbon source and soybean curd as a nitrogen source.

The pH was controlled and shifted as described above.. With a shortened culture period of 12 days, elevated activities of 530 U/ml and 8 U/ml were obtained for CMCase and p-glucosidase respectively, with an elevated protein content of 40 mg/ml in the culture broth. The soluble protein content of the culture broth was 53 mg/ml, which was considered favourable.

Table : Cellulase Productivity of* T. reesei *Mutants in Batch Culture

Mutant	*CMCase (U/ml)*	*FPU (U/ml)*	*ß-Glucosidase (U/ml)*	*Extracellular protein (mg/ml)*
QM-941 4	60	6.1	3.2	10
KY-746	88	7.2	3.5	11
K-14	106	12	1.1	17
KDR-27	150	18	2.9	20
KDG-3	324	19.5	7.7	23
PC-3-7	345	21	6.7	24
PCD-10	385	22.4	6.4	28
CDU-11	330	19	17.4	23

5-L jar fermentor, 6% Avicel, 28/C, 7 days

Table : Cellulase Productivity of T. reesei Mutants in Semi-Batch Culture

Strain	Protein (mg/ml)	CMCase (mg/ml)	ß-Glucosidase (U/ml)	Avicelase (U/ml)	FPU (U/ml)
KDG-3	36.7	448	7.6	27.6	23.2
PC-3-7	37.8	418	7.0	28.7	33.5
CDU-11	41.4	451	17.6	31.2	31.6

pH: 4.0 (adjusted with ammonia)

Culture conditions: temperature: 28/C, agitation: 550 r.p.m.; aeration: 0.8 v.v.m.

Table : Cellulase Productivity of T. reesei Mutants Using the pH Shifting System

Strain	Protein (mg/ml)	CMCase (U/ml)	ß-Glucosidase (U/ml)	Avicelase (U/ml)	FPU (U/ml)
KDG-3	22.8	332	8.7	20.1	19.5
PC-3-7	24.0	345	12.3	23.6	22.5
CDU-11	23.6	328	17.6	22.0	19.1

pH: Initial: 4.0; 5.0 after 48 hours

Culture period: 7 daysCulture conditions: temperature: $28/C_2$ agitation: 550 r.p.m.; aeration: 0.8 v.v.m.

Cellulase Production at Low Cost

Cellulase is an inducible enzyme. Sophorose, a decomposition product of crystalline cellulose, is thought to be an indispensable inducer of cellulase activity. Although cellulase culture has been investigated with the use of Avicel and other crystalline celluloses as carbon sources, cheaper carbon and nitrogen sources are desired in order to lower production costs.

An evaluation of the use of cheese whey, bagasse, and rice straw, as potential substrates for cellulase production, confirmed that PC-3-7 was capable of fermenting these substrates and efficiently producing cellulase enzymes. PC-3-7 was created as a strain, capable of partially producing cellulase enzymes. Alkali pre-treatment of biomass was observed to enhance cellulase production. This seems to suggest that cellulase production is directly proportional to the crystallinity of the biomass from which it is produced, i.e. the higher the crystallinity, the better the yield of cellulase.

The use of soybean curd, a low-cost nitrogen source, which is applicable as a substitute for expensive yeast extract and polypeptone, was investigated with cellulase fermentation residues (mainly waste cells). Results obtained, revealed that addition of 5% of the fermentation residue induced cellulase activity similar to that induced by the use of either yeast extract or polypeptone.

Table 3-5 Cellulase Production From Cheese Whey

CMCase [U/ml]	545
FPU [U/ml]	32
(3-Glucosidase [U/ml]	23
Avicelase [U/ml]	42
Protein [mg/ml]	45

Carbon source: 10% Cheese whey and 1% Avicel

Nitrogen source: Soybean meal

Potential for Mass Production of Cellulase

Cellulase fermentation was conducted with high-titer enzyme-producing strain CDU-11 under semi-batch conditions in a 4-L fermentor, using inexpensive carbon and nitrogen sources and the pH-shift system. This resulted in the production of CMCase activity exceeding 600 U/ml and (3-glucosidase activity exceeding 30 U/ml. Such activity levels could feasibly be practically applied to the development of scaled-up processes.

Pre-treatment of Cellulosic Waste

Lignocellulosic biomass contains cellulose, hemicellulose, lignin, and ash combined in a complex structure. Pre-treatment increases the crystallinity of cellulose, while removing lignin and other inhibitors, thereby enabling its enzymatic degradation. In addition, pretreatment may increase the surface area of the cellulose thereby enhancing its reactivity with the enzyme and thus its transformation.

Cellulosic biomass	*Cellulose*	*Hemicellulose*	*Lignin*	*Ash*
Bagasse	41	24	18	2
Rice straw	35	35	6	8
Wood	40-55	20-35	25-30	0.2-2.0
		(%)		

This raw material was crudely crushed (to less than 9 mm in diameter), and qualitatively fed into a counter-current extractor in which delignification was effected through counter-current contact of the raw material with sodium hydroxide, followed by washing with warm water. High alkali concentrations in the reaction mixture however caused simultaneous dissolution of pentosan, leading to a reduction in sugar recovery. The optimum biomass to sodium hydroxide ratio for favourable saccharification was determined to be 1.0:0.12. The xylose content of the saccharified solution decreased with increasing alkali concentration, under temperature conditions of 80 to 100°C, over a 1.5 to 2 hour period. Steam explosion of beech and cedar was conducted, since alkali treatment of these woods after crushing did not result in saccharification.

Saccharification of Cellulosic Waste

Saccharification

Saccharification of cellulosic waste materials was conducted using mutant strains obtained from *Trichoderma reesei* QM-9414 by mutation, as sources of cellulase enzymes. A high titer, low cost cellulase was obtained in our laboratory subsequent to extensive studies on optimum culturing of these strains.

Saccharification reaction processes were studied using three reactor types: batch, continuous stirred tank reactor (CSTR), and the tubular plug flow process. The plug flow process produced the most favourable results in that large quantities of free enzyme were recoverable from this process at late stages of the reaction.

Utilizing this process, a saccharification rate exceeding 80% was achieved at a high saccharification concentration. The adsorption of large quantities of enzyme by saccharification residues, lowered the rate of sugar production. Sugar recovery was however facilitated by stirring.

The saccharification rate and the load on the subsequent ultrafiltration (UF) and reverse osmosis (RO) membrane processes as well as energy requirements of the process, need to be carefully considered in the selection of an appropriate saccharification process. On the basis of currently available data, these processes have both benefits and disadvantages.

Table 3-7 Cellulase Requirements For The Saccharification of Bagasse

Bagasse Concentration [%]	*Saccharification Time [hours]*						
	1	2	3	4	5	6	24
1	30*	30	20	20	10	10	10
3	100	100	40	30	30	30	20
5	200	200	100	100	100	100	30
7	-	-	200	200	200	100	50
10		-	-	-	200	150	50
15	-	-	-	-	-	200	75
20	-	-	-	-	-	200	150

Enzyme: CMCase [U/ml]

* :Expressed as the enzyme titer required to achieve 80% saccharification

Recovery and Re-use of Cellulase

Although a fairly efficient cellulase production process providing a high titer cellulase at relatively low cost was developed, cellulase production was still a costly venture. The use of techniques which enable recovery and reuse of cellulase through separation of the sugar solution and cellulase from the saccharification reaction mixture, were therefore investigated using a UF membrane..

Saccharified solutions of various concentrations were centrifuged, following which unreacted residue was separated, and the filtrate was subjected to UF membrane filtration. The UF membrane filtrate was subsequently concentrated, and the resulting retentate was concentrated and re-used as recovered enzyme. During this procedure, the pressure of the UF membrane was maintained at approximately 3 kg/cm^{2}G, while the temperature of the circulating liquid was maintained between 30 and 40°C. The enzyme solution was concentrated to approximately one-tenth of its original volume. Only small amounts of residue were obtained after centrifugation of saccharified solutions of low-concentration, with sugar recovery exceeding 95%. However, when high-concentration saccharified solutions were employed, large amounts of residue resulted, and sugar recovery subsequent to centrifugation and membrane filtration was somewhat lower, *i.e.,* 83 to 89% after centrifugation and 88% subsequent to UF membrane filtration. The addition of water either to the residue or to the concentrate improved sugar recovery to a level similar to that obtained in the case of low-concentration saccharified solutions. Similar behaviour was observed with respect to enzyme recovery, although the recovery rate for the enzyme was approximately 80%. Recovered enzyme showed normal saccharification activity. UF membranes have now been successfully used for almost four years, without any degradation. These membranes can be safely operated with periodic chemical or ball-washing.

Sugar Concentration using Reverse Osmosis

Sugar solutions obtained via saccharification and enzyme recovery using the UF membrane, varied in concentration from 3 to 15%, depending upon the method used in their preparation. When subjected to alcohol fermentation, these sugar solutions produced low alcohol concentrations thus requiring large amounts of energy for alcohol concentration and dehydration.

Concentration of sugar solutions to between 25 and 30%, using a RO membrane was studied. RO was conducted by the loose RO method, using a spiral polyamide membrane, at a pressure of 40 to 50 kg/cm G and circulating liquid temperature of 25 to 30°C. In order to prevent contamination of the membrane system, warm-water washing and washing with 0.1% sodium hydroxide were performed every three days. Sugar recovery exceeding 95% was confirmed at a permeation rate of 100 to 300 L/m/d. Permeation of sugar into the filtrate was not detected.

A 25 to 30% sugar concentration was obtained regardless of the sugar concentration of the initial saccharified solution.

Table Concentration of Sugar Solutions Using RO Membrane Technology

Test run No.	*1*	*2*	*3*	*4*
Volume fed (liter)	9,180	4,840	11,460	1,900
Volume of Concentrate (liter)	600	320	400	630
Sugar Concentration (%)	26	25	28	23
Permeation Flow Velocity (liter/m²/day)	293	130	187	189

4

Microbiological Issues in Food Products Sanitation

Quality Assurance and Sanitation Programs in Food Industry

Since the late 1970s, the food industry has emphasized an organized sanitation program that monitors the microbiology of raw ingredients in production plants and the wholesomeness and safety of the finished products, in an effort to maintain or upgrade the acceptability of its food products.

As consumers become better informed and more sophisticated, it is even more vital for the food industry to develop an effective quality assurance (QA) and sanitation program. The efforts of regulatory agencies in the field of sanitation and food microbiology have been responsible for the food industry's implementation of voluntary quality assurance programs. Food scientists have also had a positive impact on quality assurance programs because many of these professionals have joined various companies in the food industry. Their efforts have been instrumental in the adoption and/or upgrading of quality assurance programs for the organizations they represent.

In its initial stages, quality assurance was primarily a quality control function, acting as an arm of manufacturing. It has now evolved to formidable force within executive structure of large food firms and has emerged into broad spectrum of activities. A quality assurance program provides the avenue to establish checks and balances in the areas of food safety, public health, technical expertise, and legal matters affecting food manufacturing firms. Activities related to food sanitation include sanitation inspections, products releases and holds, packaging sanitation, and product recalls and withdrawals.

A quality assurance program that emphasizes sanitation is vital to the growth of a food establishment. If foods are to compete effectively in the market place, established hygienic standards must be strictly maintained. However, it is sometimes impractical for production personnel to measure and monitor sanitation while maintaining a high level of productivity and efficiency. Thus, an effective quality assurance program should be available to monitor, within established priorities, each phase of the operation. All personnel should incorporate the team concept to attain established sanitary standard, ensuring that food products in the market place are safe.

Food Safety – SSOP (Standard Sanitation Operating Procedures)

Without a properly designed system to maintain sanitation, all other systems, including HACCP, will fail. It is this reason that all working meat establishments must have an active and approved SSOP program that has been signed and dated.

The SSOP consists of step-by-step procedures explaining exactly how sanitation will be maintained within the food establishment. The SSOP basically works as a plan for preventing contamination of product – derived from equipment or employee's product handling practices. The main objective of the SSOP is to allow a wholesome, safe product to be produced within a sanitary environment.

SSOP Planning and Design

The most important components of a properly developed SSOP are the Design and Execution. Each SSOP is "plant specific," meaning it was designed for a specific establishment. It is possible to modify a SSOP that has already been designed, or one can be created specifically in style of the establishment in which it will be used. As long as it meets regulatory requirements, is approved, and has been proven to be a working plan, it should be acceptable.

When deciding on the SSOP design, a general idea of the design and operations of the establishment must be known. For example, one would want to focus more on sanitation in a room that packs fully cooked product (such as bologna or sandwich meat) than that of a storage room. The storage room must be monitored, but probably not as much as the room that packs fully cooked product. Other equipment such as saws, grinders, table tops, cubing machines, knifes, and sterilizers, are examples of equipment that would need to be included in the first step of the SSOP design. There must be cleaning steps identifying how these food-contact utensils are cleaned and sanitized.

The SSOP Outline

The written SSOP should begin with a description of the establishments

SSOP program. This description should give a clear and brief understanding of the establishments standard sanitation operating procedures, including: the establishment name, city and state in which the establishment is located, an approximate number of employees working within the establishment, any special safeguards (such as good manufacturing practices and pest control programs), management structures, equipment and facilities, and a recordkeeping system.

The next step will be to describe the pre-operational sanitation check (before operations begin) where designated employees will actually perform a visual check and cleaning techniques of the equipment and facilities – placing primary focus on food-contact surfaces, and the operational sanitation check (after operations begin) where sanitation is monitored and documented during processing hours.

The SSOP "checks" are normally performed at least twice a day. These checks include a "pre-operational sanitation check" (performed before operations begin) and an "operational check" (performed during operations). Each sanitation check is performed by the designated employee and their findings are documented within a SSOP recordkeeping system. It is this recordkeeping system, as well as the actual monitoring of the SSOP, that is reviewed by USDA Inspection Personnel to verify the establishment is providing a clean and sanitary environment to produce their product.

The Pre-Operational Sanitation Check (Pre-Op)

The pre-operational sanitation check (also known as the "pre-op") is performed prior to operations. Before any food product enters a room, the pre-op should be performed in that room to verify that facilities, equipment, and utensils were properly cleaned after the previous work day. As mentioned above, all equipment and utensils are cleaned and sanitized before operations begin.

The pre-op often consists of a general walk-through of the establishment to verify the facilities, all equipment, utensils, and all food contact surfaces that may come in direct or indirect contact with the product are clean and acceptable. Often, plant personnel will use a pump sprayer, equipped with an approved sanitizer to spray all equipment and utensils before operations begin. This will likely eliminate most of any bacterial growth that may have occurred between the clean-up and the pre-op.

The Operational Sanitation Check

The operational sanitation check is performed similar to the pre-operational sanitation check, but it is performed during operations (while product is being processed), and at least once per day. The primary objective of the operational sanitation check is to verify that processing is being performed under sanitary conditions (including employee handling) to prevent any direct or cross-contamination of food products. This includes food contact surfaces as well as non-food contact surfaces. As with the pre-operational sanitation check, the results or findings of the operational sanitation check must be documented on some form of SSOP report.

SSOP Forms and Recordkeeping

It is entirely up to the establishment as to how their SSOP forms and record keeping systems are designed. The forms will be acceptable as long as they show documentation of all findings (or no findings), and any corrective actions that were needed upon a noted deficiency within the SSOP. There are no specific standards as to how the forms are designed.

Final Thoughts – Standard Sanitation Operating Procedures

Perform the sanitation checks exactly as described in the approved SSOP. Properly and accurately record any findings (or no findings if none) and complete and properly file the form(s) at the proper times. While SSOP Checks are being performed, it is recommended to use a flashlight and have the proper form(s) ready so they can be completed at the proper times. Double check the "hard-to-get" places for meat and fat debris – if these are overlooked the meat inspector will likely find them!

Identify and record any corrective actions as listed in the plant SSOP for any non-food contact surface deficiencies. Inspection Personnel will gladly assist in answering any questions that may arise pertaining to the SSOP. If there is anything that is not understood, ask the USDA Inspector. They will assist through outreach.

Food-borne Microbes and Food Poisoning

Bacteria

Of all the microorganisms, bacteria are the greatest threat to food safety. Bacteria are single-celled, living organisms that can grow quickly at favourable temperatures. However, some bacteria are useful e.g. these that to make foods like cheese, buttermilk, sauerkraut, pickles, and yogurt. Other bacteria are infectious disease-causing agents called pathogens, that use the nutrients found in potentially hazardous foods to multiply.

Some bacteria are not infectious on their own, but when they multiply in potentially hazardous food, they eject toxins that poison humans when the food is eaten.

Food handling practices are risky when they allow harmful bacteria to contaminate and grow in food. If you touch a potentially hazardous food during preparation, you may transfer several thousand bacteria to its surface.

Under the right conditions, bacteria can double every 10 to 30 minutes.

A single bacterium will double with each division—two become four, four become eight, and so on. A single cell can become billions in 10 to 12 hours.

Salmonella

Foods Associated with Salmonella: Raw poultry products, eggs, pork, processed meats. Less commonly, Salmonella has been found to be associated with raw fruits and vegetables.

Cause of illness: Infection with Salmonella species.

Incubation period: 12-24 hours.

Characteristics of Illness: Fever, cramps, diarrhea and sometimes vomiting.

Possible contaminant: Meat, poultry, egg or milk products.

Onset: Illness may begin between 7 hrs to 3 days after eating contaminated food.

Duration: Illness may last 2-3 days.;

Prevention of Illness:

- Avoid cross-contamination of ready-to-eat foods with raw meats or their juices.
- Thoroughly cook meat and poultry.
- Cook eggs thoroughly and never eat runny yolks or raw eggs.
- Always refrigerate processed meat products.
- Wash fruits and vegetables thoroughly.

Campylobacter

Foods Associated with Campylobacter: Raw chicken and raw milk.

Cause of illness: Infection, even with low numbers.

Incubation period: One to seven days.

Characteristics of Illness: Diarrhea, often associated with fever, abdominal pain, nausea, headache and muscle pain. Illness can appear very similar to Salmonellosis.

Possible contaminant: Raw milk, eggs, poultry, raw beef, cake icing, water

Onset: Illness may begin between 2-5 days after eating contaminated food.

Duration: Illness may last 7-10 days.

Staphylococcus Aureus

Foods Associated with Staphylococcus aureus: This bacteria has been associated with a wide range of foods, including meat and meat products, poultry and egg products, salads such as egg, tuna, potato and macaroni, cream-filled bakery products and pies, sandwich fillings and milk and dairy products. In general, Staph poisoning often occurs when a food has been handled a great deal (such as the chopping and handling involved in making a salad or sandwich) and is then left at temperatures above refrigeration which allow the bacteria to multiply and produce toxin.

Cause of illness: Toxin produced by certain strains of Staphylococcus aureus:

Incubation period: One-six hours.

Characteristics of Illness: Vomiting, diarrhea and abdominal cramps.

Onset: Illness may begin within 3-8 hrs. after eating contaminated food.

Duration: Illness usually lasts about 2 days.

Prevention of Illness:

- Always wash hands well when preparing foods.
- Keep foods refrigerated.

Listeria Monocytogenes

Foods Associated with Listeria monocytogenes: Raw milk, raw meats and raw vegetables. Ice cream, soft-ripened cheeses, smoked fish, lunch meats, hot dogs and refrigerated salad-type products. This organism is unique in that it is able to grow even at refrigerated temperatures and so, while refrigeration of foods will slow the growth of Listeria, it will not stop it completely.

Cause of illness: Infection with Listeria monocytogenes.

Characteristics of Illness: In healthy individuals this organism may result in diarrhea, vomiting and nausea. However, in immunocompromised individuals (the very young, the elderly, pregnant women, those with AIDS or undergoing cancer treatment) Listeriosis may first appear as mild flu-like symptoms, but may then be followed by septicemia, meningitis, encephalitis and spontaneous abortion or stillbirth in pregnant women.

Possible contaminant: Vegetables, milk, cheese, meat, seafood.

Onset: Illness may occur anywhere from 12 hrs to a few weeks after contaminated food is consumed.

Duration: In otherwise healthy individuals, mild symptoms may disappear in a day or two, but medical attention is required for immunosuppressed individuals who develop the above mentioned complications.

Prevention of Illness:

- Wash fresh fruits and vegetables thoroughly under running water.
- Keep foods refrigerated to slow the growth of Listeria, if it is present.
- Immunocompromised individuals should try to avoid eating implicated foods, such as soft-ripened cheeses or lunchmeat products.
- Pasteurize milk, cook foods properly, avoid cross- contamination; use sanitary practices.

E. Coli 0157:H7

Foods Associated with E. coli O157:H7: Undercooked raw meat, apple cider, raw milk and raw produce.

Cause of illness: Strain of enteropathic E. coli.

Incubation period: Two to four days.

Characteristics of Illness: Severe cramping and diarrhea which is initially watery but becomes grossly bloody.

Possible contaminant: Ground beef, raw milk, alfalfa sprouts, unpasteurized fruit juices, dry-cured salami, lettuce, meat, and cheese curds.

Onset: Illness may occur anywhere between 1 to 10 days after eating contaminated food, but usually occurs between 3 to 4 days.

Prevention of Illness:

- Cook ground meat thoroughly such that the to an internal reaches minimum temp of 75°C.
- Wash fresh fruits and vegetables thoroughly with running water.
- Do not drink unpasteurized milk.
- Do not drink unpasteurized apple cider unless it is well refrigerated.

Clostridium Botulinum

Foods Associated with Clostridium botulinum: Improperly processed home canned foods. and Low-acid canned foods, meats, sausage, fish.

Cause of illness: Toxin produced by Clostridium botulinum.

Incubation period: 12 to 36 hours.

Characteristics of Illness: Symptoms include double vision, vertigo, inability to swallow, Nausea, vomiting, diarrhea, fatigue, headache, dry mouth, speech difficulty and progressive respiratory paralysis or respiratory failure. Nausea may also be present initially.

Onset: Symptoms may begin within 18 hrs to 2 days of ingesting the toxin.

Duration: Greater than 65% of cases are fatal. In non-fatal botulism poisoning, recovery may take weeks to years, depending upon the severity of the poisoning.

Clostridium Perfringens

Foods Associated with Clostridium perfringens: Meat and/or gravy dishes are most often associated with this type of food borne illness. Generally C. perfringens poisoning occurs when such meat dishes are not cooked to high enough temperatures and then are allowed to sit out at room temperature for serving for an extended period of time.

Cause of illness: Contamination of drinking water or milk.

Incubation period:8 to 24 hours.

Characteristics of Illness: Abdominal cramping and diarrhea, headache, chills.

Possible contaminant: Meat, poultry, and other foods held for serving at warm, but not hot, temperatures

Onset: Illness may begin between 8 to 22 hrs after ingesting contaminated food.

Duration: Illness may last 1-2 days.

Bacillus Cereus

Foods Associated with Bacillus cereus: Rice and grain products, dairy products such as milk, cream, custards and dried milk.

Characteristics of Illness: Bacillus cereus may cause two different types of illness. The first is know as the Diarrheal illness, which results in diarrhea and abdominal cramps occurring within 6 to 15 hrs of eating contaminated food. This illness may persist up to about 24 hours and resembles Clostridium perfringens food poisoning. The second type of illness caused by Bacillus cereus is know as the Emetic illness and results in nausea and vomiting within 3-6 hours of eating contaminated food. This illness also lasts about 24 hrs and it tends to resemble Staphylococcal food poisoning.

There are two types of food poisoning: infectious agent and toxic agent. Food infection refers to the presence of bacteria or other microbes which infect the body after consumption. Food intoxication refers to the ingestion of toxins contained within the food, including bacterially produced exotoxins, which can happen even when the microbe that produced the toxin is no longer present or able to cause infection. In spite of the common term food poisoning, most cases are caused by a variety of pathogenic bacteria, viruses, or parasites that contaminate food, rather than chemical or natural toxins.

Foodborne illness usually arises from improper handling, preparation, or food storage. Good hygiene practices before, during, and after food preparation can reduce the chances of contracting an illness. There is a general consensus in the public health community that regular hand-washing is one of the most effective defences against the spread of foodborne illness. The action of monitoring food to ensure that it will not cause foodborne illness is known as food safety. Foodborne disease can also be caused by a large variety of toxins that affect the environment. For foodborne illness caused by chemicals.

Foodborne illness can also be caused by pesticides or medicines in food and naturally toxic substances like poisonous mushrooms or reef fish.

Bacteria

Bacteria are a common cause of foodborne illness. In the United Kingdom during 2000 the individual bacteria involved were as follows: *Campylobacter jejuni* 77.3%, *Salmonella* 20.9%, *Escherichia coli O157:H7* 1.4%, and all others less than 0.1%. In the past, bacterial infections were thought to be more prevalent because few places had the capability to test for norovirus and no active surveillance was being done for this particular agent. Symptoms for bacterial infections are delayed because the bacteria need time to multiply. They are usually not seen until 12–72 hours or more after eating contaminated food.

Most common bacterial foodborne pathogens are:

- *Campylobacter jejuni* which can lead to secondary Guillain–Barre syndrome and periodontitis
- *Clostridium perfringens*, the "cafeteria germ"
- *Salmonella* spp. – its *S. typhimurium* infection is caused by consumption of eggs or poultry that are not adequately cooked or by other interactive human-animal pathogens
- *Escherichia coli O157:H7* enterohemorrhagic (EHEC) which causes hemolytic-uremic syndrome.

Other common bacterial foodborne pathogens are:

- *Bacillus cereus*
- *Escherichia coli*, other virulence properties, such as enteroinvasive (EIEC), enteropathogenic (EPEC), enterotoxigenic (ETEC), enteroaggregative (EAEC or EAgEC)
- *Listeria monocytogenes*

- *Shigella* spp.
- *Staphylococcus aureus*
- *Streptococcus*
- *Vibrio cholerae,* including O1 and non-O1
- *Vibrio parahaemolyticus*
- *Vibrio vulnificus*
- *Yersinia enterocolitica* and *Yersinia pseudotuberculosis.*

Less common bacterial agents:

- *Brucella* spp.
- *Corynebacterium ulcerans*
- *Coxiella burnetii* or Q fever
- *Plesiomonas shigelloides.*

Exotoxins

In addition to disease caused by direct bacterial infection, some foodborne illnesses are caused by exotoxins which are excreted by the cell as the bacterium grows. Exotoxins can produce illness even when the microbes that produced them have been killed. Symptoms typically appear after 1–6 hours depending on the amount of toxin ingested.

- *Clostridium botulinum*
- *Clostridium perfringens*
- *Staphylococcus aureus*
- *Bacillus cereus.*

For example *Staphylococcus aureus* produces a toxin that causes intense vomiting. The rare but potentially deadly disease botulism occurs when the anaerobic bacterium *Clostridium botulinum* grows in improperly canned low-acid foods and produces botulin, a powerful paralytic toxin.

Pseudoalteromonas tetraodonis, certain species of Pseudomonas and Vibrio, and some other bacteria, produce the lethal tetrodotoxin, which is present in the tissues of some living animal species rather than being a product of decomposition.

Mycotoxins and Alimentary Mycotoxicoses

The term alimentary mycotoxicoses refers to the effect of poisoning by Mycotoxins through food consumption. Mycotoxins sometimes have important effects on human and animal health. For example, an outbreak which occurred in the UK in 1960 caused the death of 100,000 turkeys which had consumed aflatoxin-contaminated peanut meal. In the USSR in World War II, 5000 people died due to Alimentary Toxic Aleukia (ALA). The common foodborne Mycotoxins include:

- Aflatoxins – originated from Aspergillus parasiticus and Aspergillus flavus. They are frequently found in tree nuts, peanuts, maize, sorghum and other oilseeds, including corn and cottonseeds. The pronounced forms of Aflatoxins are those of B1, B2, G1, and G2, amongst which Aflatoxin B1 predominantly targets the liver, which will result in necrosis, cirrhosis, and carcinoma. In the US, the acceptable level of total aflatoxins in foods is less than 20 μg/kg, except for Aflatoxin M1 in milk, which should be less than 0.5 μg/kg. The official document can be found at FDA's website.
- Altertoxins – are those of Alternariol (AOH), Alternariol methyl ether (AME), Altenuene (ALT), Altertoxin-1 (ATX-1), Tenuazonic acid (TeA) and Radicinin (RAD), originated from Alternaria spp. Some of the toxins can be present in sorghum, ragi, wheat and tomatoes. Some research has shown that the toxins can be easily cross-contaminated between grain commodities, suggesting that manufacturing and storage of grain commodities is a critical practice.
- Citrinin
- Citreoviridin
- Cyclopiazonic acid
- Cytochalasins
- Ergot alkaloids/Ergopeptine alkaloids – Ergotamine
- Fumonisins – Crop corn can be easily contaminated by the fungi Fusarium moniliforme, and its Fumonisin B1 will cause Leukoencephalomalacia (LEM) in horses, Pulmonary edema syndrome (PES) in pigs, liver cancer in rats and Esophageal cancer in humans. For human and animal health, both the FDA and the EC have regulated the content levels of toxins in food and animal feed.
- Fusaric acid
- Fusarochromanone
- Kojic acid
- Lolitrem alkaloids
- Moniliformin
- 3-Nitropropionic acid
- Nivalenol
- Ochratoxins – In Australia, The Limit of Reporting (LOR) level for Ochratoxin A (OTA) analyses in 20th Australian Total Diet Survey was 1 μg/kg, whereas the EC restricts the content of OTA to 5 μg/

kg in cereal commodities, 3 µg/kg in processed products and 10 µg/kg in dried vine fruits.

- Oosporeine
- Patulin – Currently, this toxin has been advisably regulated on fruit products. The EC and the FDA have limited it to under 50 µg/kg for fruit juice and fruit nectar, while limits of 25 µg/kg for solid-contained fruit products and 10 µg/kg for baby foods were specified by the EC.
- Phomopsins
- Sporidesmin A
- Sterigmatocystin
- Tremorgenic mycotoxins – Five of them have been reported to be associated with molds found in fermented meats. These are Fumitremorgen B, Paxilline, Penitrem A, Verrucosidin, and Verruculogen.
- Trichothecenes – sourced from Cephalosporium, Fusarium, Myrothecium, Stachybotrys and Trichoderma. The toxins are usually found in molded maize, wheat, corn, peanuts and rice, or animal feed of hay and straw. Four trichothecenes, T-2 toxin, HT-2 toxin, diacetoxyscirpenol (DAS) and deoxynivalenol (DON) have been most commonly encountered by humans and animals. The consequences of oral intake of, or dermal exposure to, the toxins will result in Alimentary toxic aleukia, neutropenia, aplastic anemia, thrombocytopenia and/or skin irritation. In 1993, the FDA issued a document for the content limits of DON in food and animal feed at an advisory level. In 2003, US published a patent that is very promising for farmers to produce a trichothecene-resistant crop.
- Zearalenone
- Zearalenols.

Emerging Foodborne Pathogens

Many foodborne illnesses remain poorly understood. Approximately sixty percent of outbreaks are caused by unknown sources.

- *Aeromonas hydrophila, Aeromonas caviae, Aeromonas sobria.*

Preventing Bacterial Food Poisoning

Prevention is mainly the role of the state, through the definition of strict rules of hygiene and a public services of veterinary surveying of animal products in the food chain, from farming to the transformation industry and delivery (shops and restaurants). This regulation includes:

- traceability: in a final product, it must be possible to know the origin of the ingredients (originating farm, identification of the harvesting or of the animal) and where and when it was processed; the origin of the illness can thus be tracked and solved (and possibly penalized), and the final products can be removed from the sale if a problem is detected;

- enforcement of hygiene procedures like HACCP and the "cold chain";
- power of control and of law enforcement of veterinarians.

In August 2006, the United States Food and Drug Administration approved Phage therapy which involves spraying meat with viruses that infect bacteria, and thus preventing infection. This has raised concerns, because without mandatory labelling consumers wouldn't be aware that meat and poultry products have been treated with the spray.

At home, prevention mainly consists of good food safety practices. Many forms of bacterial poisoning can be prevented even if food is contaminated by cooking it sufficiently, and either eating it quickly or refrigerating it effectively. Many toxins, however, are not destroyed by heat treatment.

Viruses

Viral infections make up perhaps one third of cases of food poisoning in developed countries. In the US, more than 50% of cases are viral and noroviruses are the most common foodborne illness, causing 57% of outbreaks in 2004. Foodborne viral infection are usually of intermediate (1–3 days) incubation period, causing illnesses which are self-limited in otherwise healthy individuals, and are similar to the bacterial forms described above.

- Enterovirus
- Hepatitis A is distinguished from other viral causes by its prolonged (2–6 week) incubation period and its ability to spread beyond the stomach and intestines, into the liver. It often induces jaundice, or yellowing of the skin, and rarely leads to chronic liver dysfunction. The virus has been found to cause the infection due to the consumption of fresh-cut produce which has fecal contamination.
- Hepatitis E
- Norovirus
- Rotavirus.

Parasites

Most foodborne parasites are zoonoses.

- Platyhelminthes:

- o *Diphyllobothrium* sp.
- o *Nanophyetus* sp.
- o *Taenia saginata*
- o *Taenia solium*
- o *Fasciola hepatica.*

- Nematode:
  - o *Anisakis* sp.
  - o *Ascaris lumbricoides*
  - o *Eustrongylides* sp.
  - o *Trichinella spiralis*
  - o *Trichuris trichiura.*
- Protozoa:
  - o *Acanthamoeba* and other free-living amoebae
  - o *Cryptosporidium parvum*
  - o *Cyclospora cayetanensis*
  - o *Entamoeba histolytica*
  - o *Giardia lamblia*
  - o *Sarcocystis hominis*
  - o *Sarcocystis suihominis*
  - o *Toxoplasma gondii.*

Natural Toxins

Several foods can naturally contain toxins, many of which are not produced by bacteria. Plants in particular may be toxic; animals which are naturally poisonous to eat are rare. In evolutionary terms, animals can escape being eaten by fleeing; plants can use only passive defences such as poisons and distasteful substances, for example capsaicin in chili peppers and pungent sulfur compounds in garlic and onions. Most animal poisons are not synthesised by the animal, but acquired by eating poisonous plants to which the animal is immune, or by bacterial action.

- Alkaloids
- Ciguatera poisoning
- Grayanotoxin (honey intoxication)
- Mushroom toxins
- Phytohaemagglutinin (red kidney bean poisoning; destroyed by boiling)
- Pyrrolizidine alkaloids

- Shellfish toxin, including paralytic shellfish poisoning, diarrhetic shellfish poisoning, neurotoxic shellfish poisoning, amnesic shellfish poisoning and ciguatera fish poisoning
- Scombrotoxin
- Tetrodotoxin (fugu fish poisoning).

Some plants contain substances which are toxic in large doses, but have therapeutic properties in appropriate dosages.

- Foxglove contains cardiac glycosides.
- Poisonous hemlock (conium) has medicinal uses.

Other Pathogenic Agents

- Prions, resulting in Creutzfeldt-Jakob disease.

"Ptomaine Poisoning"

An early theory on the causes of food poisoning involved *ptomaines* (from Greek *ptôma*, "fall, fallen body, corpse"), alkaloids found in decaying animal and vegetable matter. While some alkaloids do cause poisoning, the discovery of bacteria left the ptomaine theory obsolete and the word *ptomaine* is no longer used scientifically.

Mechanism

Incubation Period

The delay between consumption of a contaminated food and appearance of the first symptoms of illness is called the incubation period. This ranges from hours to days (and rarely months or even years, such as in the case of Listeriosis or Creutzfeldt-Jacob disease), depending on the agent, and on how much was consumed. If symptoms occur within 1–6 hours after eating the food, it suggests that it is caused by a bacterial toxin or a chemical rather than live bacteria.

The long incubation period of many foodborne illnesses tends to cause sufferers to attribute their symptoms to "stomach flu".

During the incubation period, microbes pass through the stomach into the intestine, attach to the cells lining the intestinal walls, and begin to multiply there. Some types of microbes stay in the intestine, some produce a toxin that is absorbed into the bloodstream, and some can directly invade the deeper body tissues. The symptoms produced depend on the type of microbe.

Infectious Dose

The infectious dose is the amount of agent that must be consumed to give rise to symptoms of foodborne illness, and varies according to the agent and the consumer's age and overall health. In the case of *Salmonella* a relatively large inoculum of 1 million to 1 billion organisms is necessary to produce symptoms in healthy human volunteers, as *Salmonellae* are very sensitive to acid. An unusually high stomach pH level (low acidity) greatly reduces the number of bacteria required to cause symptoms by a factor of between 10 and 100.

Epidemiology

Every year there are an estimated 76 million foodborne illnesses in the United States (26,000 cases for 100,000 inhabitants), 2 million in the United Kingdom (3,400 cases for 100,000 inhabitants) and 750,000 in France (1,220 cases for 100,000 inhabitants).

United States

In the United States, using FoodNet data from 1996–1998, the CDCP estimated there were 76 million foodborne illnesses (26,000 cases for 100,000 inhabitants):

- 325,000 were hospitalized (111 per 100,000 inhabitants);
- 5,000 people died (1.7 per 100,000 inhabitants.).
- Major pathogens from foodborne illness in the United States cost upwards of US $35 billion in medical costs and lost productivity (1997).

France

In France, for 750,000 cases(1,210 per 100,000 inhabitants):

- 70,000 people consulted in the emergency department of an hospital (113 per 100,000 inhabitants.);
- 113,000 people were hospitalized (24 per 100,000 inhabitants);
- 400 people died (0.9 per 100,000 inhabitants).

Australia

In Australia, there are an estimated 5.4 million cases of food-borne illness every year, causing:

- 18,000 hospitalizations
- 120 deaths
- 2.1 million lost days off work
- 1.2 million doctor consultations
- 300,000 prescriptions for antibiotics.

Outbreaks

The vast majority of reported cases of foodborne illness occur as individual or sporadic cases. The origin of most sporadic cases is undetermined. In the United States, where people eat outside the home frequently, most outbreaks (58%) originate from commercial food facilities (2004 FoodNet data). An outbreak is defined as occurring when two or more people experience similar illness after consuming food from a common source.

Often, a combination of events contributes to an outbreak, for example, food might be left at room temperature for many hours, allowing bacteria to multiply which is compounded by inadequate cooking which results in a failure to kill the dangerously elevated bacterial levels.

Outbreaks are usually identified when those affected know each other. However, more and more, outbreaks are identified by public health staff from unexpected increases in laboratory results for certain strains of bacteria. Outbreak detection and investigation in the United States is primarily handled by local health jurisdictions and is inconsistent from district to district. It is estimated that 1–2% of outbreaks are detected.

Pathogens and Antibiotics

A pathogen an infectious agent, or more commonly germ, is a biological agent that causes disease to its host. There are several substrates including *pathways* whereby pathogens can invade a host; the principal pathways have different episodic time frames, but soil contamination has the longest or most persistent potential for harboring a pathogen.

The body contains many natural orders of defence against some of the common pathogens (such as *Pneumocystis*) in the form of the human immune system and by some "helpful" bacteria present in the human body's normal flora. However, if the immune system or "good" bacteria is damaged in any way (such as by chemotherapy, human immunodeficiency virus (HIV), or antibiotics being taken to kill other pathogens), pathogenic bacteria that were being held at bay can proliferate and cause harm to the host. Such cases are called opportunistic infection.

Some pathogens (such as the bacterium *Yersinia pestis* which may have caused the Black Plague, the *Variola* virus, and the Malaria protozoa) have been responsible for massive numbers of casualties and have had numerous effects on afflicted groups. Of particular note in modern times is HIV, which is known to have infected several million humans globally, along with the Influenza virus. Today, while many medical advances have been made to safeguard against infection by pathogens, through the use of vaccination, antibiotics, and fungicide, pathogens continue to threaten human life. Social advances such as food safety, hygiene, and water treatment have reduced the threat from some pathogens. Not all pathogens are negative. In entomology, pathogens are one of the "Three P's" (predators, pathogens, and parasitoids) that serve as natural or introduced biological controls to suppress arthropod pest populations.

Types of Pathogen

Viral

Pathogenic viruses are mainly those of the families of: Adenoviridae, Picornaviridae, Herpesviridae, Hepadnaviridae, Flaviviridae, Retroviridae, Orthomyxoviridae, Paramyxoviridae, Papovaviridae, Polyomavirus, Rhabdoviridae, Togaviridae. Some notable pathogenic viruses cause: smallpox, influenza, mumps, measles, chickenpox, ebola, and rubella. Viruses typically range between 20-300 nanometers in length.

Bacterial

Although the vast majority of bacteria are harmless or beneficial, a few pathogenic bacteria can cause infectious diseases. The most common bacterial disease is tuberculosis, caused by the bacterium *Mycobacterium tuberculosis*, which affects about 2 million people mostly in sub-Saharan Africa.

Pathogenic bacteria contribute to other globally important diseases, such as pneumonia, which can be caused by bacteria such as *Streptococcus* and *Pseudomonas*, and foodborne illnesses, which can be caused by bacteria such as *Shigella*, *Campylobacter* and *Salmonella*. Pathogenic bacteria also cause infections such as tetanus, typhoid fever, diphtheria, syphilis and Hansen's disease. Bacteria can often be killed by antibiotics. They typically range between 1-5 micrometers in length.

Fungal

Fungi comprise a eukaryotic kingdom of microbes that are usually saprophytes but can cause diseases in humans, animals and plants. Fungi are the most common cause of diseases in crops and other plants. Life threatening fungal infections in humans most often occur in immunocompromised patients or vulnerable people with a weakened immune system, although fungi are common problems in the immunocompetent population as the causative agents of skin, nail or yeast infections. Most antibiotics that function on bacterial pathogens cannot be used to treat fungal infections due to the fact that fungi and their hosts both have eukaryotic cells. Most clinical fungicides belong to the azole group. The typical fungal spore size is 1-40 micrometer in length.

Other Parasites

Some eukaryotic organisms, such as protists and helminths, cause disease. One of the best known diseases caused by protists in the genus *Plasmodium* is malaria. These can range from 3-200 micrometers in length.

Prionic

Prions are infectious pathogens that do not contain nucleic acids. Prions are abnormal proteins whose presence causes some diseases such as scrapie, bovine spongiform encephalopathy (mad cow disease) and Creutzfeldt–Jakob disease.. The discovery of prion as a new class of pathogen has lead Stanley B. Prusiner to receive Nobel Prize in Physiology or Medicine in 1997.

Potency

One hypothesis regarding pathogens states that the longer a pathogen can survive outside of the body, the more dangerous it can be to a potential host. For example, the smallpox virus (*variola virus*) can survive outside the human body for approximately 885 days. It is also one of the most deadly pathogenic viruses, as it kills between 20-50% of the people it infects. The tuberculosis bacterium kills 1 in 5 of the people it infects, but only survives 244 days outside of its host. However, research into the basis of the ability of pathogens to cause disease provides evidence from multiple and diverse species of the existence of pathogenicity or virulence factors, encoded within the pathogens' genetic material, that facilitate microbes to cause disease.

In countries that have higher sanitation standards, pathogens cannot survive for as long outside of the human. This is seen as encouragement to mutations to the pathogen which would make it less deadly, as such mutations would allow the pathogen to survive in the host for longer periods of time.

Transmission

One of the primary pathways by which food or water become contaminated is from the release of untreated sewage into a drinking water supply or onto cropland, with the result that people who eat or drink contaminated sources become infected. In developing countries most sewage is discharged into the environment or on cropland; even in developed countries there are periodic system failures resulting in a sanitary sewer overflow.

Antimicrobial

An anti-microbial is a substance that kills or inhibits the growth of microorganisms such as bacteria, fungi, or protozoans. Antimicrobial drugs either kill microbes (microbiocidal) or prevent the growth of microbes (microbiostatic). Disinfectants are antimicrobial substances used on non-living objects or outside the body.

The history of antimicrobials begins with the observations of Pasteur and Joubert, who discovered that one type of bacteria could prevent the growth of another. They did not know at that time that the reason one bacterium failed to grow was that the other bacterium was producing an antibiotic. Technically, antibiotics are only those substances that are produced by one microorganism that kill, or prevent the growth, of another microorganism. Of course, in today's common usage, the term antibiotic is used to refer to almost any drug that attempts to rid your body of a bacterial infection. Antimicrobials include not just antibiotics, but synthetically formed compounds as well.

The discovery of antimicrobials like penicillin and tetracycline paved the way for better health for millions around the world. Before penicillin became a viable medical treatment in the early 1940s, no true cure for gonorrhea, strep throat, or pneumonia existed. Patients with infected wounds often had to have a wounded limb removed, or face death from infection. Now, most of these infections can be cured easily with a short course of antimicrobials.

However, with the development of antimicrobials, microorganisms have adapted and become resistant to previous antimicrobial agents. The old antimicrobial technology was based either on poisons or heavy metals, which may not have killed the microbe completely, allowing the microbe survive, change, and become resistant to the poisons and/or heavy metals.

Antimicrobial nanotechnology is a recent addition to the fight against disease causing organisms, replacing heavy metals and toxins and may some day be a viable alternative. Infections that are acquired during a hospital visit are called "hospital acquired infections" or nosocomial infections. Similarly, when the infectious disease is picked up in the non-hospital setting it is considered "community acquired".

Main Classes

Antibiotics

Antibiotics are generally used to treat bacterial infections. The toxicity to humans and other animals from antibiotics is generally considered to be low. However, prolonged use of certain antibiotics can decrease the number of gut flora, which can have a negative impact on health. Some recommend that, during or after prolonged antibiotic use, one should consume probiotics and eat reasonably to replace destroyed gut flora.

The term antibiotic originally described only those formulations derived from living organisms, but is now applied also to synthetic antimicrobials, such as the sulfonamides.

The discovery, development, and clinical use of antibiotics during the 20th century has decreased substantially the mortality from bacterial infections. The antibiotic era began with the pneumatic application of nitroglycerine drugs, followed by a "golden" period of discovery from about

1945 to 1970, when a number of structurally diverse, highly effective agents were discovered and developed. However, since 1980 the introduction of new antimicrobial agents for clinical use has declined, in part because of the enormous expense of developing and testing new drugs. Paralleled to this there has been an alarming increase in bacterial resistance to existing agents.

Antibiotics are among the most commonly used drugs. For example, 30% or more hospitalized patients are treated with one or more courses of antibiotic therapy. However, antibiotics are also among the drugs commonly misused by physicians, e.g. usage of antibiotic agents in viral respiratory tract infections. The inevitable consequence of widespread and injudicious use of antibiotics has been the emergence of antibiotic-resistant pathogens, resulting in a serious threat to global public health. The resistance problem demands that a renewed effort be made to seek antibacterial agents effective against pathogenic bacteria resistant to current antibiotics. One of the possible strategies towards this objective is the rational localization of bioactive phytochemicals.

Antivirals

Antiviral drugs are a class of medication used specifically for treating viral infections. Like antibiotics for bacteria, specific antivirals are used for specific viruses. Unlike most antibiotics, antiviral drugs do not destroy their target pathogen; instead they inhibit their development.

Antiviral drugs are one class of antimicrobials, a larger group which also includes antibiotic, antifungal and antiparasitic drugs. They are relatively harmless to the host, and therefore can be used to treat infections. They should be distinguished from viricides, which are not medication but destroy virus particles.

Most of the antivirals now available are designed to help deal with HIV, herpes viruses (best known for causing cold sores and genital herpes, but actually causing a wide range of diseases), the hepatitis B and C viruses, which can cause liver cancer, and influenza A and B viruses. Researchers are working to extend the range of antivirals to other families of pathogens.

Designing safe and effective antiviral drugs is difficult, because viruses use the host's cells to replicate. This makes it difficult to find targets for the drug that would interfere with the virus without also harming the host organism's cells.

The emergence of antivirals is the product of a greatly expanded knowledge of the genetic and molecular function of organisms, allowing biomedical researchers to understand the structure and function of viruses, major advances in the techniques for finding new drugs, and the intense pressure placed on the medical profession to deal with the human immunodeficiency virus (HIV), the cause of the deadly acquired immunodeficiency syndrome (AIDS) pandemic.

Almost all anti-microbials, including anti-virals, are subject to drug resistance as the pathogens mutate over time, becoming less susceptible to the treatment. For instance, a recent study published in Nature Biotechnology emphasized the urgent need for augmentation of oseltamivir (Tamiflu) stockpiles with additional antiviral drugs including zanamivir (Relenza) based on an evaluation of the performance of these drugs in the scenario that the 2009 H1N1 'Swine Flu' neuraminidase (NA) were to acquire the tamiflu-resistance (His274Tyr) mutation which is currently widespread in seasonal H1N1 strains.

History

Through the mid- to late-20th century, medical science and practice included an array of effective tools, ranging from antiseptics to vaccines and antibiotics, but no drugs to treat viral infections. While vaccines were effective in preventing many viral diseases, they could not help once a viral infection set in. Prior to the development of antivirals, when someone contracted a virus, there was little that could be done other than treating the symptoms and waiting for the disease to run its course.

The first experimental antivirals were developed in the 1960s, mostly to deal with herpes viruses, and were found using traditional trial-and-error drug discovery methods. Researchers grew cultures of cells and infected them with the target virus. They then introduced chemicals into the cultures they thought were likely to inhibit viral activity, and observed whether the level of virus in the cultures rose or fell. Chemicals that seemed to have an effect were selected for closer study. This was a very time-consuming, hit-or-miss procedure, and in the absence of a good knowledge of how the target virus worked, it was not efficient in discovering antivirals that were effective and had few side effects. It was not until the 1980s, when the full genetic sequences of viruses began to be unraveled, that researchers began to learn how viruses worked in detail, and exactly what chemicals were needed to thwart their reproductive cycle. Dozens of antiviral treatments are now available, and medical research is rapidly exploiting new knowledge and technology to develop more.

Virus Life Cycle

Viruses consist of a genome and sometimes a few enzymes stored in a capsule made of protein (called a capsid), and sometimes covered with a lipid layer (sometimes called an 'envelope'). Viruses cannot reproduce on their own, and instead propagate by subjugating a host cell to produce copies of themselves, thus producing the next generation.

Researchers working on such "rational drug design" strategies for developing antivirals have tried to attack viruses at every stage of their life cycles. Some species of mushrooms have been found to contain multiple antiviral chemicals with similar synergistic effects. Viral life cycles vary in their precise details depending on the species of virus, but they all share a general pattern:

- Attachment to a host cell.
- Release of viral genes and possibly enzymes into the host cell.
- Replication of viral components using host-cell machinery.
- Assembly of viral components into complete viral particles.
- Release of viral particles to infect new host cells.

Limitations of Vaccines

Vaccines bolster the body's immune system to better attack viruses in the "complete particle" stage, outside of the organism's cells. They traditionally consist of an attenuated (weakened or killed) version of the virus. These vaccines can, in rare cases, harm the host by inadvertently infecting the host with a full-blown viral occupancy. Recently "subunit" vaccines have been devised that consist strictly of protein targets from the pathogen. They stimulate the immune system without doing serious harm to the host. In either case, when the real pathogen attacks the subject, the immune system responds to it quickly and blocks it.

Vaccines are very effective on stable viruses, but are of limited use in treating a patient who has already been infected. They are also difficult to successfully deploy against rapidly mutating viruses, such as influenza (the vaccine for which is updated every year) and HIV. Antiviral drugs are particularly useful in these cases.

Anti-viral Targeting

The general idea behind modern antiviral drug design is to identify viral proteins, or parts of proteins, that can be disabled. These "targets" should generally be as unlike any proteins or parts of proteins in humans as possible, to reduce the likelihood of side effects. The targets should also be common across many strains of a virus, or even among different species of virus in the same family, so a single drug will have broad effectiveness. For example, a researcher might target a critical enzyme synthesized by the virus, but not the patient, that is common across strains, and see what can be done to interfere with its operation.

Once targets are identified, candidate drugs can be selected, either from drugs already known to have appropriate effects, or by actually designing the candidate at the molecular level with a computer-aided design program.

The target proteins can be manufactured in the lab for testing with candidate treatments by inserting the gene that synthesizes the target protein into bacteria or other kinds of cells. The cells are then cultured for mass production of the protein, which can then be exposed to various treatment candidates and evaluated with "rapid screening" technologies.

Approaches by Life Cycle Stage

Before Cell Entry

One anti-viral strategy is to interfere with the ability of a virus to infiltrate a target cell. The virus must go through a sequence of steps to do this, beginning with binding to a specific "receptor" molecule on the surface of the host cell and ending with the virus "uncoating" inside the cell and releasing its contents. Viruses that have a lipid envelope must also fuse their envelope with the target cell, or with a vesicle that transports them into the cell, before they can uncoat.

This stage of viral replication can be inhibited in two ways:

- 1. Using agents which mimic the virus-associated protein (VAP) and bind to the cellular receptors. This may include VAP anti-idiotypic antibodies, natural ligands of the receptor and anti-receptor antibodies.
- 2. Using agents which mimic the cellular receptor and bind to the VAP. This includes anti-VAP antibodies, receptor anti-idiotypic antibodies, extraneous receptor and synthetic receptor mimics.

This strategy of designing drugs can be very expensive, and since the process of generating anti-idiotypic antibodies is partly trial and error, it can be a relatively slow process until an adequate molecule is produced.

Entry Inhibitor

A very early stage of viral infection is viral entry, when the virus attaches to and enters the host cell. A number of "entry-inhibiting" or "entry-blocking" drugs are being developed to fight HIV. HIV most heavily targets the immune system's white blood cells known as "helper T cells", and identifies these target cells through T-cell surface receptors designated "CD4" and "CCR5". Attempts to interfere with the binding of HIV with the CD4 receptor have failed to stop HIV from infecting helper T cells, but research continues on trying to interfere with the binding of HIV to the CCR5 receptor in hopes that it will be more effective.

Uncoating Inhibitor

Inhibitors of uncoating have also been investigated.

Amantadine and rimantadine, have been introduced to combat influenza. These agents act on penetration/uncoating.

Pleconaril works against rhinoviruses, which cause the common cold, by blocking a pocket on the surface of the virus that controls the uncoating process. This pocket is similar in most strains of rhinoviruses and enteroviruses, which can cause diarrhea, meningitis, conjunctivitis, and encephalitis.

During Viral Synthesis

A second approach is to target the processes that synthesize virus components after a virus invades a cell.

Reverse Transcription

One way of doing this is to develop nucleotide or nucleoside analogues that look like the building blocks of RNA or DNA, but deactivate the enzymes that synthesize the RNA or DNA once the analogue is incorporated. This approach is more commonly associated with the inhibition of reverse transcriptase (RNA to DNA) than with "normal" transcriptase (DNA to RNA).

The first successful antiviral, acyclovir, is a nucleoside analogue, and is effective against herpesvirus infections. The first antiviral drug to be approved for treating HIV, zidovudine (AZT), is also a nucleoside analogue.

An improved knowledge of the action of reverse transcriptase has led to better nucleoside analogues to treat HIV infections. One of these drugs, lamivudine, has been approved to treat hepatitis B, which uses reverse transcriptase as part of its replication process. Researchers have gone further and developed inhibitors that do not look like nucleosides, but can still block reverse transcriptase.

Another target being considered for HIV antivirals include RNase H- which is a component of reverse transcriptase that splits the synthesized DNA from the original viral RNA.

Integrase

Another target is integrase, which splices the synthesized DNA into the host cell genome.

Transcription

Once a virus genome becomes operational in a host cell, it then generates messenger RNA (mRNA) molecules that direct the synthesis of viral proteins. Production of mRNA is initiated by proteins known as transcription factors. Several antivirals are now being designed to block attachment of transcription factors to viral DNA.

Translation/Antisense

Genomics has not only helped find targets for many antivirals, it has provided the basis for an entirely new type of drug, based on "antisense" molecules. These are segments of DNA or RNA that are designed as complementary molecule to critical sections of viral genomes, and the binding of these antisense segments to these target sections blocks the operation of those genomes. A phosphorothioate antisense drug named fomivirsen has been introduced, used to treat opportunistic eye infections in AIDS patients caused by cytomegalovirus, and other antisense antivirals are in development. An antisense structural type that has proven especially valuable in research is morpholino antisense.

Morpholino oligos have been used to experimentally suppress many viral types:

- caliciviruses
- flaviviruses (including WNV)
- dengue
- HCV
- coronaviruses.

Translation/Ribozymes

Yet another antiviral technique inspired by genomics is a set of drugs based on ribozymes, which are enzymes that will cut apart viral RNA or DNA at selected sites. In their natural course, ribozymes are used as part of the viral manufacturing sequence, but these synthetic ribozymes are designed to cut RNA and DNA at sites that will disable them. A ribozyme antiviral to deal with hepatitis C has been suggested, and ribozyme antivirals are being developed to deal with HIV. An interesting variation of this idea is the use of genetically modified cells that can produce custom-tailored ribozymes. This is part of a broader effort to create genetically modified cells that can be injected into a host to attack pathogens by generating specialized proteins that block viral replication at various phases of the viral life cycle.

Protease Inhibitors

Some viruses include an enzyme known as a protease that cuts viral protein chains apart so they can be assembled into their final configuration. HIV includes a protease, and so considerable research has been performed to find "protease inhibitors" to attack HIV at that phase of its life cycle. Protease inhibitors became available in the 1980s and have proven effective, though they can have unusual side effects, for example causing fat to build up in unusual places. Improved protease inhibitors are now in development. Protease inhibitors have also been seen in nature. A protease inhibitor was isolated from the Shiitake mushroom (*Lentinus edodes*). The presence of this may explain the Shiitake mushrooms noted antiviral activity *in vitro*.

Assembly

Rifampicin acts at the assembly phase.

Release Phase

The final stage in the life cycle of a virus is the release of completed viruses from the host cell, and this step has also been targeted by antiviral drug developers. Two drugs named zanamivir (Relenza) and oseltamivir (Tamiflu) that have been recently introduced to treat influenza prevent the release of viral particles by blocking a molecule named neuraminidase that is found on the surface of flu viruses, and also seems to be constant across a wide range of flu strains.

Immune System Stimulation

A second category of tactics for fighting viruses involves encouraging the body's immune system to attack them, rather than attacking them directly. Some antivirals of this sort do not focus on a specific pathogen, instead stimulating the immune system to attack a range of pathogens.

One of the best-known of this class of drugs are interferons, which inhibit viral synthesis in infected cells. One form of human interferon named "interferon alpha" is well-established as part of the standard treatment for hepatitis B and C, and other interferons are also being investigated as treatments for various diseases.

A more specific approach is to synthesize antibodies, protein molecules that can bind to a pathogen and mark it for attack by other elements of the immune system. Once researchers identify a particular target on the pathogen, they can synthesize quantities of identical "monoclonal" antibodies to link up that target. A monoclonal drug is now being sold to help fight respiratory syncytial virus in babies, and antibodies purified from infected individuals are also used as a treatment for hepatitis B.

Antifungals

An antifungal drug is medication used to treat fungal infections such as athlete's foot, ringworm, candidiasis (thrush), serious systemic infections such as cryptococcal meningitis, and others.

Antifungals work by exploiting differences between mammalian and fungal cells to kill off the fungal organism without dangerous effects on the host. Unlike bacteria, both fungi and humans are eukaryotes. Thus, fungal and human cells are similar at the molecular level, making it more difficult to find a target for an antifungal drug to attack that does not also exist in the infected organism. Consequently, there are often side effects to some of these drugs. Some of these side effects can be life-threatening if the drug is not used properly.

Antimicrobials and Home Mold remediation- Anti-fungal treatments are frequently sought-after to treat mold growth in damp or wetted home materials that exhibit mold growth. Note that most home mold problems are moisture/water-caused and the solution for conquering the mold growth is most dependent upon the water/moisture control and removal/discarding of the mold-damaged materials. Cleaning mold-damaged materials may result in a visually acceptable appearance but most cleaning methods kill mold or prevent its return. For this reason, moisture management is the primary focus for mold prevention. Generally, Relative Humidity levels in the home above 54% will support mold groth on most cellulose containing materials (fabrics, carpeting and carpet backing, wood, paper, boxes, dust and lint). Mold also readily grows on most latex paints and leather.

Antimicrobials used in home remediation follow a variety of chemistry and functions. One common method of mold remediation utilizes Sodium Bicarbonate (Baking soda) as a blasting medium much in the way that sand is used to blast (clean) surfaces under the propulsion of compressed air. This 'Soda Blasting' makes a large cloud of dust that is pH Alkalai and its residue that permeates the wood and painted surfaces is also naturally antimicrobial as a result of the high pH and the presence of Sodium Bicarbonate. If used alone, repeated wetting can wash away the Sodium Bicarbonate residue and mold can return to the materials if the water source is not managed. Dry ice (frozen Carbon Dioxide (CO2)) is also used as a blasting agent where clean-up is more restrictive (i.e. attics) and while the dry ice blasting leaves no antimicrobial residue, it does prep the blasted surface to receive one as a secondary step. One popular, professional anti-microbial (Serum) is often applied after or without blasting by soda or dry ice and in one of its variants, is a mix of Hydrogen Peroxide and thin surface coating that neutralizes the mold (making it non-viable) and encapsulating the surface to prevent spore release. Other anti-microbial surface treatments typically contain variants of metals known to supress mold grwoth; i.e. pigments or solutions involving Copper, Silver, Zinc or other metals (some of which can be toxic to humans if improperly applied). Most antimicrobial solutions are professionally applied and are not sold to the public.

Many people use either white vinegar or laundry bleach as an inexpensive anti-microbial solution. These liquids are best used in combination and they can be safely combined. When combined in equal amounts (i.e. 2 cups bleach + 2 cups vinegar in 2 gallons of warm water + 2 drops of dish soap), these make a solution known as "Acidified Bleach" and it is considerably more effective as an antimicroboial and as a disinfectant. On porous foundation surfaces, a secondary step of scrubbing with Tri-Sodium Phosphate (TSP) or Spic-N-Span is capable of leaving behind a phosphate or carbonate reside that will be antimicrobial. The key to successful foundation cleaning with longer-lasting results hinges on leaving the cleaner residue on the wall and force-drying the area with a fan to drive off the water. Long-term success requires that the water source be corrected and that the area's relative humiidity be kept as far below 54% as possible.

Antimicrobials and paints- Kitchen and Bath paint formulations are often manufactured with the understanding that these areas are often cleaned and may experience elevated humidity levels from bathing/cooking. As a result, leading manufacturers often design Kitchen and Bath formulations to be less porous, more scrubable, and often, the paint formulation is adjusted to be more antimicrobial than other interior paints. During colour mixing, some stores also offer the option of buying and adding an antimicrobial packet to the paint as a booster. It is ill-advised to use a normal latex paint on a foundation since this naturally moist area will readily cause mold to grow on latex. Foundation painting should be done with products designed for that purpose and when applied over virgin masonry, these products (i.e. DryLok by United Gilsonite Labs) perform very well.

Antiparasitics

Antiparasitics are a class of medications which are indicated for the treatment of infection by parasites, such as nematodes, cestodes, trematodes, infectious protozoa, and amoebae. Like antifungals, they must kill the infecting pest without serious damage to the host.

Non-pharmaceutical Antimicrobials

A wide range of chemical and natural compounds are used as antimicrobials. Organic acids are used widely as antimicrobials in food products, e.g. lactic acid, citric acid, acetic acid, and their salts, either as ingredients, or as disinfectants. For example, beef carcasses often are sprayed with acids, and then rinsed or steamed, to reduce the prevalence of E. coli O157:H7.

Traditional healers long have used plants to prevent or cure infectious disease. Many of these plants have been investigated scientifically for antimicrobial activity, and a large number of plant products have been shown to inhibit the growth of pathogenic microorganisms. A number of these agents appear to have structures and modes of action that are distinct from those of the antibiotics in current use, suggesting that cross-resistance with agents already in use may be minimal. So, it is worthwhile to study plants and plant products for activity against resistant bacteria.

Essential Oils

The antimicrobial properties of 21 plant essential oils and two essences were investigated against five food-borne pathogens, *Campylobacter jejuni, Salmonella enteriditis, Escherichia coli, Staphylococus aureus and Listeria monocytogenes*. The oils of bay, cinnamon, clove and thyme were the most inhibitory, each having a bacteriostatic concentration of 0.075% or less against all five pathogens. (A. Smith-Palmer, J. Stewart and L. Fyfe. Antimicrobial properties of plant essential oils and essences against five important food-borne pathogens. Letters in Applied Microbiology 1998. 26. 118-122)

Many essential oils are included in pharmacopoeias as having antimicrobial activity, including:

- Sideritis or Greek Mountain Tea
- Oregano oil
- Tea tree oil- in cosmetics, medicine
- Mint oil- in medicine, cosmetics (tooth paste etc.)
- Sandalwood oil- in cosmetics
- Clove oil- stomatology etc.
- Nigella sativa (Black cumin) oil
- Onion oil (*Allium cepe*)- phytoncides, in phytotherapy
- Leleshwa oil
- Lavender oil
- Lemon oil
- Lemon myrtle oil
- Neem oil
- Garlic
- Eucalyptus oil
- Peppermint oil
- Cinnamon oil
- Thyme oil.

Cations and Elements

Many heavy metal cations such as Hg^{2+}, Cu^{2+}, and Pb^{2+} have antimicrobial activities, but are also very toxic to other living organisms, thus making them unsuitable for treating infectious diseases. Colloidal silver is commonly used as an antimicrobial in alternative medicine without clear scientific proof of effectiveness.

Important Antibiotics and their Applications

Antibiotics are substances that inhibit the growth of microorganisms (anti- metabolites) or their replication (a bacteriostatic effect). They were traditionally obtained by extracting them from cultures of microbes. However, most drugs on the market today are semisynthetic derivatives of natural products. Sulfa drugs, discovered in the 1980s, were the first antimicrobial agents put into clinical use. Unfortunately, many bacteria are not susceptible to sulfonamides, and with the outbreak of World War II came the need for other more potent antibacterial agents. The serendipitous discovery of penicillin is, without a doubt, the most celebrated breakthrough in the history of antibiotics. In the late 1920s, while working in a London hospital, Alexander Fleming observed a mold overtaking a culture of staphylococcus bacteria he was growing in his laboratory. He extracted juices from the mold and, in 1929, reported that the extract, which he called penicillin, had antiseptic (anti-infectious) activity. The fungus was subsequently identified as *Penicillium notatum* (now called *Penicillium chrysogenum).* It was not until the 1940s that penicillin was put into clinical use. Howard W. Florey, professor of pathology at Oxford's Sir William Dunn School of Pathology, and Ernst B. Chain are credited with culturing the fungus and producing the first significant quantities of penicillin for treating bacterial infections. In 1945 Fleming, Florey, and Chain received the Nobel Prize in physiology or medicine "for the discovery of penicillin and its curative effect in various infectious diseases."

Extracts from microorganisms are still an important source of antibiotics today. Clinically, antibiotics are described as possessing either broad- or narrow-spectrum activity. Bacteria are classified based on a staining technique developed by Danish microbiologist Hans Christian Gram. The bacterial cell walls of gram-positive bacteria stain blue when treated with either crystal violet or methylene blue, while gram-negative bacteria do not retain the stain and appear red. Broad-spectrum antibiotics are capable of inhibiting both gram-positive and gram-negative bacterial cultures. Grampositive bacteria have simpler cell walls than gram-negative strains and are susceptible to less toxic, narrow-spectrum antibiotics.

β-Lactam Antibiotics

A variety of penicillins have been produced by the fermentation of *Penicillium chrysogenum* in the presence of different nutrients. Penicillin G predominates when the culture medium is rich in phenylacetic acid, whereas the incorporation of phenoxyacetic acid favours penicillin V (phenoxymethylpenicillin). Semisynthetic penicillins, such as ampicillin and amoxicillin, are prepared by replacing the aromatic side chain of biosynthetically derived penicillins with other chemical groups. All penicillins are β-lactam antibiotics and have the same mechanism of action: They inhibit bacterial cell wall biosynthesis.

Bacterial cell walls differ from mammalian cell walls and are therefore attractive targets for antibiotics. Bacterial cell walls contain β-lactam receptors, known as penicillin-binding proteins (PBPs). β-Lactam antibiotics bind to the PBPs of bacterial cell walls and prevent their growth and repair. Widespread use of penicillin, however, has led to drug resistance. Because microorganisms multiply rapidly, strains of bacteria with enzymes capable of hydrolyzing β-lactam rings (β-lactamases) have evolved. β-lactamases are capable of inactivating β-lactam antibiotics before they bind to receptors on cell walls. As a result, physicians sometimes prescribe β-lactamase inhibitors to patients on penicillin therapy to circumvent drug inactivation by bacterial enzymes.

The cephalosporins comprise another important class of broad-spectrum β-lactam antibiotics. Cephalosporins were originally isolated from cultures of *Cephalosporium acremonium.* Cephalexin (Keflex) is a semisynthetic cephalosporin frequently prescribed to treat ear and skin infections caused by staphylococci or streptococci.

Antibiotics that Inhibit Protein Synthesis

There are also a large number of antibiotics structurally unrelated to penicillins and cephalosporins. These compounds exert their antimicrobial activity by inhibiting protein biosynthesis. In 1947 chloramphenicol was isolated from cultures of *Streptomyces venezuelae.* It is a broad-spectrum bacteriostatic agent that interferes with protein synthesis by binding to bacterial ribosomes. The use of chloramphenicol in humans is limited because of the drug's toxicity. It inhibits liver enzymes and suppresses red blood cell formation.

Aminoglycosides are amino sugars with broad-spectrum antibiotic activity. Streptomycin, isolated from *Streptomyces griseus,* was the first aminoglycoside antibiotic discovered. Although streptomycin initially proved to be a potent agent against gram-negative bacteria, rapid microbial resistance to the drug has limited its use and today streptomycin is generally administered in combination with other antibiotics. Neomycin is a broad-spectrum aminoglycoside antibiotic isolated from *Streptomyces fradiae.* However, because of the adverse effects of neomycin on the kidneys and ear, its use in humans is restricted to topical applications, often in combination with other antibiotics or corticosteroids. Concerns over the potential risks associated with aminoglycoside therapy, chiefly nephrotoxicity (kidney disease) and ototoxicity (damage to the ear canal), have diminished their use.

In 1952 the broad-spectrum antibiotic erythromycin was isolated from cultures of *Streptomyces erythreus* (later renamed *Saccharopolyspora erythraea).* The erythromycins are macrolide antibiotics that typically have a 12-, 14-, or 16-membered cyclic backbone which is a lactone.

Erythromycin A, the major fermentation component of *S. erythraea*, is a 14-membered ring macrolide that is used by medicinal chemists as the foundation for building semisynthetic derivatives of erythromycin antibiotics. (Macrolides inhibit bacteria by interfering with microbial protein biosynthesis.) Semisynthetic macrolides are popular with clinicians because they can be administered orally and have relatively low toxicity. They are often used to treat respiratory tract infections, and have been especially effective against conditions such as Legionnaires' disease and communityacquired pneumonia. Erythromycin therapy is often prescribed for individuals allergic to penicillin. One of the most widely used macrolide antibiotics derived from erythromycin A is azithromycin (Zithromax). Resistance to macrolide antibiotics generally involves mutations of bacterial ribosomal RNA that prevent macrolide binding.

In the late 1940s and early 1950s a series of tetracycline antibiotics was isolated from cultures of streptomyces. All tetracyclines consist of four fused 6-membered rings. Tetracyclines are broad-spectrum antibiotics that interfere with protein synthesis by binding to bacterial ribosomes. Unfortunately, the frequent use of tetracyclines to treat minor infections has led to resistance among previously susceptible strains of bacteria (pneumococci and staphylococci). Resistance to tetracyclines occurs when bacteria either develop proteins that prevent ribosomal binding by tetracyclines, or synthesize enzymes capable of inactivating tetracyclines.

Widespread use of antibiotics and rapid microbial evolution have led to highly resistant bacterial strains. Although most scientists no longer believe that a single drug will be developed to wipe out all infectious diseases, there is increasing demand for new antimicrobial agents. Currently, combined drug therapy appears to be the most effective means of circumventing microbial resistance to antibiotics.

P Enicillin

Penicillin was discovered accidentally in 1929 when Sir Alexander Fleming observed bacterial cultures contaminated with a mold that inhibited bacterial growth. The antibiotic penicillin was subsequently isolated from cultures of the *Penicillium* mold. In 1938 two other British scientists, Howard

Florey and Ernst Chain, first used purified preparations of penicillin to treat bacterial infections. Penicillin may have been present in folk remedies used as early as 600 B.C.E., at around which time molded soybean curd was used by the Chinese to treat boils and carbuncles, and moldy cheese was used by Chinese and Ukrainian peasants to treat infected wounds.

Initially it was thought that penicillin was a pure substance, but further studies revealed that a number of closely related compounds were present in *Penicillium* cultures. Naturally occurring penicillins, such as penicillin G, are most effective against gram-positive bacteria, but much less effective against gram-negative bacteria. A further limitation to the use of Penicillin G is that it is not well absorbed when administered orally. Research programs to produce chemically modified penicillins with improved properties have resulted in a large number of clinically useful penicillin derivatives. Examples of such penicillin derivatives include ampicillin and amoxicillin, which have much greater efficacy against gram-negative bacteria than penicillin, retain good activity against gram-positive bacteria, and are well absorbed when administered orally. The principal adverse reaction associated with the penicillins is the occurrence of allergic response.

The molecular targets for the antibacterial activity of the penicillin and related β-lactam antibiotics such as the cephalosporins are a group of bacterial enzymes known as *penicillin-binding proteins* (PBPs). The PBPs are essential to the final stages of bacterial cell wall biosynthesis. Penicillin and other β-lactam antibiotics inhibit PBPs, thereby inhibiting bacterial cell wall biosynthesis, which eventually results in bacterial cell lysis. (Vancomycin and cycloserine are nonpenicillin antibiotics that also inhibit bacterial cell wall biosynthesis through other mechanisms.)

The penicillins and related antibiotics have been among the most widely used therapeutic agents since their introduction into clinical practice in the 1940s. However, the widespread use of these antibiotics has resulted in the emergence and spread of bacteria that are resistant to these agents. A major mechanism of resistance to the penicillin and other β-lactam antibiotics is the bacterial production of β-lactamases, enzymes that cleave the β-lactam antibiotics and render them inactive before they can inhibit their PBP targets. Significant efforts have been made to develop β-lactam antibiotics resistant to the β-lactamases, and toward finding inhibitors of the β-lactamases to allow β-lactam antibiotics to be useful antibacterial agents against β-lactamase producing bacteria.

Gram-positive and Gram-negative Bacteria

Bacteria can be broadly classified into two groups; the gram-positive bacteria, which are stained purple after the gram staining procedure, and the gram-negative bacteria, which are stained red. The difference in staining reflects differences in the structure of the cell walls between gram-positive and gram-negative bacteria. Pathogenic gram-positive bacteria include Staphylococcus aureus and Bacillus anthracis, and pathogenic gram-negative bacteria include Escherichia coli and Neisseria gonorrhoeae.

S Ulfa D Rugs

Sulfa drugs were the first synthetic drugs with widespread antibiotic activity to be put into clinical use. In the 1930s German chemists observed that certain dyes used to stain bacteria stopped microbial growth. Gerhard Domagk, a pathologist at I. G. Farbenindustrie, performed a series of experiments on mice infected with streptococcus bacteria and observed that mice injected with an orange-red dye called Prontosil survived bacterial infection. Prontosil is an azo dye that had not shown antibacterial activity during earlier in vitro tests. However, in vivo, the dye is transformed into sulfanilamide, a compound with antimicrobial activity. Domagk had such faith in Prontosil's anti-infectious properties that he is reported to have injected the dye into his daughter when she had septicemia. In 1939 Domagk was awarded the Nobel Prize in physiology or medicine for his discovery.

Once sulfanilamide was recognized as an active antimicrobial agent, scientists synthesized thousands of sulfonamides to test for bactericidal activity. It was later realized that sulfonamides do not actually kill bacteria; they interfere with bacterial growth and replication. Sulfa drugs are bactericstatic. They inhibit an enzyme necessary for the biosynthesis of folic acid in bacteria. Folic acid is necessary for the biosynthesis of thymine and the purine bases, the building blocks of DNA. Bacteria that are sensitive to sulfa drugs are unable to acquire folic acid from their environment and, in the presence of sulfonamides, are unable to synthesize the folic acid essential for cell growth and multiplication. Sulfonamides do not harm their human hosts, however, because, unlike susceptible bacteria, humans acquire folic acid from their diet and lack the enzyme necessary for synthesizing folic acid.

Production of Penicillin

Many fungal species grow on food products, often contributing to their spoilage, as in fruits, vegetables, and other fresh or prepared products. However, in many instances the presence of fungi on some products does not have a deleterious effect and may even contribute to the ripening and development of important organoleptic characteristics typical of a particular food (17, 19). This is the case of fungi belonging to the genus *Penicillium*, which grow on cured meat products and on many kinds of cheese. Some of these fungi whose presence is beneficial for the product are nowadays used as starters, as a means to colonize homogeneously the cheese or meat piece, thus preventing the eventual colonization by undesirable fungi and bacteria. Some of the fungal species most widely used as starters are *Penicillium nalgiovense* and *Penicillium chrysogenum* for meat products, *Penicillium roqueforti* for Roquefort cheese, and *Penicillium camemberti* for Camembert cheese. Most fungi growing on food are producers of mycotoxins and antibiotics (28, 39, 46), and the presence of the compounds secreted by the fungus in the colonized food product has been reported on different occasions (9, 31). Laich et al. (25) showed that *P. nalgiovense* produces penicillin on the surface of a Spanish fermented sausage (fuet); the presence of the antibiotic can be detected in the outer layers of the sausages. The presence of penicillin in food is not desirable, as it can produce allergic reactions (18, 20, 29) and might eventually lead to the development of resistance phenomena in human resident bacteria, which could then transfer that genetic information to pathogenic bacteria, as resistance gene transfer between bacteria in the human colon has been shown to be a common phenomenon (42). The development of resistance phenomena in bacteria as a consequence of the ingestion of subtherapeutic levels of antibiotics has been reported in several cases in farm animals (1, 34).

Therefore, it is important to know whether fungi associated with food production are potential penicillin producers or not. The objective of the present work was to analyse the possible production of penicillin and to study the presence of the penicillin biosynthetic genes in fungi of the genus *Penicillium* growing on food products or used as starters for those products.

Materials and Methods

Penicillium Strains

Some of the fungal species were chosen for the present study because they are commonly used as commercial starters (*P. nalgiovense*, *P. chrysogenum*, *P. camemberti*, and *P. roqueforti*); the rest of the species were chosen because of their frequent isolation from cured meat products like cecina (a salted and smoked-cured beef meat product from north Spain) as part of the natural mycobiota growing on this product (F. Laich, F. Fierro, and J. F. Martín, unpublished results).

P. chrysogenum NRRL 1951; *P. nalgiovense* NRRL 911; *Penicillium griseofulvum* NRRL 2300, NRRL 2152, NRRL 989, NRRL 994, NRRL 991, and NRRL 992; *P. roqueforti* NRRL 849; *P. camemberti* NRRL 874, NRRL 877, NRRL 876 (ex-type of *Penicillium candidum*), and CECT 2267; *Penicillium verrucosum* NRRL 5573 and NRRL 5574; *Penicillium brevicompactum* NRRL 862; *Penicillium commune* NRRL 845; *Penicillium solitum* NRRL 937; *Penicillium crustosum* NRRL 972; *Penicillium expansum* NRRL 976; *Penicillium implicatum* NRRL 2061; *Penicillium hirsutum* NRRL 2032; *Penicillium aurantiogriseum* NRRL 971; *Penicillium viridicatum* NRRL 963; *Penicillium echinulatum* NRRL 1151; *Penicillium purpurogenum* ATCC MYA-38; and *Paecilomyces variotii* NRRL 1775 were obtained either from the ARS Culture Collection (Peoria, Ill.), the American Type Culture Collection (ATCC) (Manassas, Va.), or the Spanish type culture collection (CECT) (Valencia, Spain).

Other strains were isolated from cheese (inner part) and cecina (surface) as indicated by Motilva Casado et al.. For colony isolation, samples were plated on malt extract agar (MEA) and incubated at 25°C for 7 days. The strains isolated were identified to species level according to the method of Pitt and Hocking and genetic analysis. Once the strains were identified they were named as follows: *P. nalgiovense* 16a, *P. commune* INBCC300, and *P. chrysogenum* AK (all isolated from cecina); and *P. roqueforti* CONT1 (isolated from blue cheese).

P. chrysogenum AS-P-78 was kindly donated by Antibióticos, S.A. (León, Spain). *P. nalgiovense* INBCC103 is a commercial starter culture.

Identification Procedures

All isolates were identified to the species level according to the method of Pitt and Hocking. To confirm this identification, the randomly amplified polymorphic DNA (RAPD) technique was used. Total DNA was obtained as described by Fierro et al. RAPD PCR amplification was carried out according tc the protocol described by Williams et al. with modifications. PCRs were performed with a 25-µl volume containing the following: 16 mM $(NH_4)_2SO_4$; 67 mM Tris-HCl, pH 8.8; 0.01% Tween 20; 3.5 mM $MgCl_2$; a 200 µM concentration (each) of dATP, dCTP, dGTP, and dTTP (Fermentas); 0.32 µM primer; 0.5 U of *Taq* DNA polymerase (Bioline); and 10 to 20 ng of genomic DNA. The following oligonucleotides were used: 1, 5'-AGTCAGCCAC-3'; 2, 5'-AATCGGGCT-3'; CRL9, 5'-CAGCCGCCCC-3'; and CRL12, 5'-CCGCCGCCCG-3'. PCR was carried out with a Whatman Biometra T Gradient thermal cycler using an initial denaturation step of 94°C for 4 min; followed by 45 cycles of 40 s at 94°C, 1 min at 34°C, and 2 min at 72°C; and a final extension of72°C for 10 min. Amplified fragments were separated on a 1.5% (wt/vol) agarose gel and visualized by ethidium bromide staining.

Bioassays for Penicillin Production

All the strains described above were tested for their ability to produce penicillin by bioassays in both solid and liquid medium with *Micrococcus luteus* ATCC 9341 as a test strain. ß-Lactamase from *Bacillus cereus* UL1 (300 µl from the supernatant of a *B. cereus* culture per 100 ml of tryptic soy agar [TSA] medium) was used to check if the antibacterial activity observed in the samples was due to the presence of penicillin or to some other antagonistic compound. The bioassay on solid medium was performed by growing the fungus in petri dishes on MEA at 25°C for 7 to 10 days (when a high number of conidia was produced by the colony). Agar plugs (9-mm diameter) were taken out of the fungal colonies, placed on the surface of TSA medium (Difco) containing 1% agar, and incubated for 48 h at 25°C. Then, an overlayer of 1% TSA with *M. luteus* was added onto the medium with the plugs, the antibiotic was allowed to diffuse in the plates in the cold (2 h at 5°C), and finally the cultures were incubated at 30°C for 24 h.

Studies of the production of penicillin on liquid cultures were carried out as described previously.

HPLC Analysis

P. griseofulvum NRRL 2300 was fermented on a liquid culture in CPM medium under the conditions described before. Samples from the supernatant were taken after 96 h of fermentation. Fifty microliters of both the unprocessed supernatant and an ethyl acetate-extracted fraction obtained as described previously by Laich et al. was analysed with a Waters high-performance liquid chromatograph (HPLC) with a μBondapak C_{18} column. Penicillin G at a concentration of 0.1 mg/ml was used as the control sample.

Buffer A was 50 mM sodium acetate, pH 4.5, and buffer B was acetonitrile. The flow rate was 1.3 ml/min, and the running conditions were as follows: min 1 to 3, buffer A; min 3 to 15, gradient buffer B, 0 to 60%; min 15 to 18, gradient buffer B, 60 to 80%; min 18 to 20, buffer B, 80%; min 20 to 22, gradient buffer B, 80 to 0%; and min 22 to 25, buffer A.

Southern Blotting and Hybridization

Total DNA of the fungi was extracted as described before. Five micrograms of DNA of each strain was digested with the restriction endonuclease *Eco*RI. The DNA fragments were separated on a horizontal 0.6% agarose gel and transferred onto a Hybond-N membrane (Amersham Pharmacia Biotech) using a vacuum blotter (VacuGene XL; Pharmacia Biotech). The probes were labelled with the DIG DNA labeling mixture (Boehringer Mannheim) according to the manufacturer's protocol. Prehybridization and hybridization were done with 40% formamide standard buffer at 42°C.

After hybridization, the membrane was washed at room temperature for 15 min with 2x SSC (1x SSC is 0.15 M NaCl plus 0.015 M sodium citrate)-0.1% sodium dodecyl sulfate (SDS), 15 min at 42°C with 0.1x SSC-0.1% SDS, and 3 min at 65°C with 0.1x SSC-0.1% SDS. The signals were visualized with a chemiluminescent substrate for alkaline phosphatase (Roche) according to the manufacturer's protocol.

PCR Amplification and Sequencing

The following primers were used for amplification by PCR: NALAB1, 5'-CTTCTCGAGGGGCTAGTCATGAGTCCGTATTCA-3'; NALAB2, 5'-CAAGGATCCTTGGAATCACACTTTCTACGTCCG-3'; PVCG, 5'-RTCRATDCKDCCSAGRTACTCRAT-3'; PVCB, 5'-GGHGGHGCYTAYGTBCCNATYGA-3'; PCAB1, 5' CTGGATATCC-TAACGACCGC-3'; and PCAB2, 5' CAGGTAGAGCTCGCCGACAG-3'.

The letter code used for nucleotide designation is according to the Nomenclature Committee of the International Union of Biochemistry.

PCR amplification was performed by standard procedures. In the case of primers PVCG and PVCB, an annealing temperature gradient between 46 and 69°C was used.

DNA sequencing was performed with an ABI PRISM 310 Genetic Analyser (PE Applied Biosystems) using standard sequencing primers and the ABI PRISM sequencing kit, according to the manufacturer's instructions.

Results

Penicillin Production by Different Strains

All the strains were first screened to detect penicillin producers using a very sensitive bioassay in solid medium. No antibacterial activity was observed after five bioassays of strains belonging to the species *P. roqueforti*, *P. camemberti*, *P. brevicompactum*, *P. commune*, *P. solitum*, *P. expansum*, *P. implicatum*, *P. hirsutum*, *P. aurantiogriseum*, *P. viridicatum*, *P. echinulatum*, *P. purpurogenum*, and *Paecilomyces variotii*. However, strains belonging to *P. chrysogenum*, *P. nalgiovense*, *P. griseofulvum*, *P. crustosum*, and *P. verrucosum* consistently showed different levels of antibacterial activity. The highest antibacterial activity was found in *P. chrysogenum* AK, followed by *P. griseofulvum* NRRL 2300 and by *P. nalgiovense* 16a. ß-Lactamase from *B. cereus* UL1 was used as a means to discern between the antibacterial activity caused by penicillin and that exerted by other antagonistic substances that the fungus might produce. However, in three other strains of *P. griseofulvum* (NRRL 991, NRRL 992, and NRRL 989) only part of the antibacterial activity could be attributed to penicillin, as a smaller penicillinase-resistant inhibition halo was still formed in the presence of ß-lactamase.

In *P. verrucosum* and *P. crustosum*, no decrease of the inhibition halo was observed upon treatment with ß-lactamase, which indicates that the antibacterial activity observed in this case is not caused by penicillin but by some other compound(s).

Penicillin production by *P. griseofulvum* strains in solid medium was confirmed by fermentation in liquid submerged cultures. *P. griseofulvum* strain NRRL 2300 showed the highest production level, while the rest of the *P. griseofulvum* strains produced much smaller amounts (less than 10 µg/ml). In some of the strains, penicillin production was detected only after 80 h of fermentation.

An HPLC analysis of the supernatant of a *P. griseofulvum* NRRL 2300 liquid culture was performed to confirm that the antibiotic produced by this fungus was penicillin. It was, therefore, concluded that *P. griseofulvum* produces authentic penicillin G.

Analysis of the Presence of Penicillin Biosynthetic Genes in Different Strains

The results on production of penicillin observed in the previous experiments should correlate with the presence or absence ofthe penicillin biosynthetic genes *pcbAB*, *pcbC*, and *penDE* in the different strains. We performed a series of Southern blot analyses to detect the possible presence of the genes in all the strains used in this study. Probes C, D, and E correspond to the *pcbAB* gene, encoding L--aminoadipyl-L-cysteinyl-D-valine (ACV) synthetase, the multidomain enzyme catalyzing the first step of the biosynthesis of penicillin. Probes A and B correspond to the *pcbC* and *penDE* genes, respectively, which encode the enzymes responsible for the second and third step of the biosynthesis. The results obtained with probes A and B indicate that both the *pcbC* and the *penDE* genes are present in all the strains of *P. chrysogenum*, *P. nalgiovense*, and *P. griseofulvum* and are absent in all the remaining strains belonging to other species. The presence of the penicillin gene cluster confirms that *P. griseofulvum* is a penicillin producer.

Similarly, the absence of the penicillin cluster in *P. verrucosum* and *P. crustosum* supports the conclusion that the antibacterial activity observed in these fungi is not due to penicillin. All the strains of *P. griseofulvum* contained the whole *pcbAB* gene, as revealed by the hybridization signals appearing with the three probes C, D, and E. The Southern blot analyses performed with these five probes, along with additional Southern blots with other probes from the cluster, allowed us to elaborate on an *Eco*RI restriction map of the new penicillin gene cluster in *P. griseofulvum* by comparison with the previously described maps of *P. chrysogenum* (6) and *P. nalgiovense* (25). There is an *Eco*RI site in the *penDE* gene of the *P. griseofulvum* NRRL 991 strain which is absent in the other strains of *P. griseofulvum*. In general most of the restriction sites are conserved between the three species, but it is striking that one of the *Eco*RI fragments inside *pcbAB* is about 160 bp larger in *P. griseofulvum* than in *P. chrysogenum*.

Production of Streptomycin

Streptomycin is an antibiotic drug, the first of a class of drugs called aminoglycosides to be discovered, and was the first antibiotic remedy for tuberculosis. It is derived from the actinobacterium *Streptomyces griseus*. Streptomycin is a bactericidal antibiotic. Streptomycin cannot be given orally, but must be administered by regular intramuscular injections. An adverse effect of this medicine is ototoxicity, which can cause ear damage.

Streptomycin is a protein synthesis inhibitor. It binds to the S12 Protein of the 30S subunit of the bacterial ribosome, interfering with the binding of formyl-methionyl-tRNA to the 30S subunit. This prevents initiation of protein synthesis and leads to death of microbial cells. Humans have structurally different ribosomes from bacteria, thereby allowing the selectivity of this antibiotic for bacteria. However at low concentrations Streptomycin only inhibits growth of the bacteria by inducing prokaryotic ribosomes to misread mRNA. Streptomycin is an antibiotic that inhibits both Gram-positive and Gram-negative bacteria, and is a therefore a useful broad spectrum antibiotic.

History

Streptomycin was first isolated on October 19, 1943 by Albert Schatz, a graduate student, in the laboratory of Selman Abraham Waksman at Rutgers University. Waksman and his laboratory discovered several antibiotics, including actinomycin, clavacin, streptothricin, streptomycin, grisein, neomycin, fradicin, candicidin and candidin. Of these, streptomycin and neomycin found extensive application in the treatment of numerous infectious diseases. Streptomycin was the first antibiotic that could be used to cure the disease tuberculosis; early production of the drug was dominated by Merck & Co. under George W. Merck.

The first randomized trial of streptomycin against pulmonary tuberculosis was carried out in 1947 by the MRC Tuberculosis Research Unit. Whilst neither double-blind nor placebo-controlled, results showed efficacy against TB, albeit with minor toxicity and acquired bacterial resistance to the drug.

Streptomycin Production

Streptomycin and various other antibiotics are produced using strains of Streptomyces griseus. Spores of this actinomycete are inoculated into a medium to establish a culture with a high mycelial biomass for introduction into an inoculum tank, with subsequent use of the mycelial inoculum to initiate the fermentation process in the production tank.

The medium contains soybean meal (N- source), glucose (C- source) and NaCl. Process is carried out at 28°C and the maximum production achieved at pH range of 7.6-8.0. High agitation and aeration are needed. Process lasts for about 10 days. The classic fermentation process involves three phases. During the first phase there is rapid growth of the microbe with production of mycelial biomass.

Proteolytic activity of the microbe releases NH3 to the medium from the soybean meal, causing a rise in pH. During this initial fermentation phase there is little production of streptomycin. During second phase there is little additional production of mycelium, but the secondary metabolite, streptomycin accumulates in the medium. The glucose and NH3 released are consumed during this phase.

The pH remains fairly constant-between 7.6 and 8.0. In the third and final phase, when carbohydrates become depleted, streptomycin production ceases and the microbial cells begin to lyse pH increases and process normally ends by this time.

After the process is complete, mycelium is separated from the broth by filtration and the antibiotic recovered. In one method of recovery and purification, streptomycin is adsorbed onto activated charcol and eluted with acid alcohol. It is then precipitated with acetone and further purified by use of column chromatography.

Uses

Treatment of Diseases

- Infective endocarditis caused by enterococcus when the organism is not sensitive to Gentamicin
- Tuberculosis in combination with other anti-TB drugs. It is not the first line treatment, except in medically under-served populations where the cost of more expensive treatments are prohibitive.
- Plague (*Yersinia pestis*) has historically been treated with it as the first line treatment. It is approved for this purpose by the U.S. Food and Drug Administration.
- In veterinary medicine, streptomycin is the first line antibiotic for use against gram negative bacteria in large animals (horses, cattle, sheep etc.). It is commonly combined with procaine penicillin for intramuscular injection.

While streptomycin is traditionally given intramuscularly (indeed, in many countries it is only licensed to be used intramuscularly), the drug may also be administered intravenously.

Pesticide

Streptomycin is also used as a pesticide, to combat the growth of bacteria, fungi, and algae. Streptomycin controls bacterial and fungal diseases of certain fruit, vegetables, seed, and ornamental crops, and controls algae in ornamental ponds and aquaria. A major use is in the control of fireblight on apple and pear trees. As in medical applications, extensive use can be associated with the development of resistant strains.

Cell Culture

Streptomycin, in combination with penicillin, is used in a standard antibiotic cocktail to prevent bacterial infection in cell culture.

Mechanism and Systemic mode of Action of Antibiotics in Plants

Fungicides can be classified by chemical group, general mode of action, specific mode of action, or by physical properties once in the plant. Many fungicides within a family, such as the benzimidazoles (group 1), have the same mode of action against fungi. Often, it is recommended to tank-mix or to alternate fungicides with different modes of action to prevent or delay the buildup of resistant fungi. In general, fungi resistant to one chemical in a group, such as iprodione, also are resistant to all other chemicals in the group- in this example of group 2 fungicides, to vinclozolin.

A fungicide's mode of action can be described in general or specific terms. A fungicide with broad-spectrum activity is effective against a large variety of pathogenic fungi. Examples of broad-spectrum fungicides include captan, sulfur, and mancozeb. Some fungicides have a very narrow spectrum of activity; for example, metalaxyl is effective only against oomycetes like *Phytophthora*. Alternatively, a fungicide may affect a broad range of fungi but by only a specific mode of action. For example, Topsin is useful to control many fungal diseases; it acts by binding to tubulin, thereby blocking mitosis.

Another way to describe a fungicide's mode of action is to say that it is used for protection, as a curative, or as an eradicant.

Protectants are applied to healthy plants to prevent fungal spores from germinating or penetrating host tissue. They must be applied before the fungal spore has a chance to infect the plant. New plant tissue that develops after application generally is unprotected. Protectants generally are not effective once the fungus grows into plant tissues. Examples of protectants include mancozeb, coppers, and chlorothalonil.

Note: Some formulations of chlorothalonil, such as Bravo, can protect newly developed plant tissues because rain action redistributes the fungicide to other plant parts.

Curative materials generally act within the plant and are effective against fungi shortly after penetration. "Kickback activity" and "curative" are interchangeable terms. These materials must be applied within a certain time after infection starts. Materials such as dodine, triflumizole, or myclobutanil have 36-, 72-, and 96-hour curative activity, respectively, against the apple scab fungus.

Eradicants can be of two types. Lime sulfur is an eradicant that acts by killing fungi on contact. For example, lime sulfur can be used during the dormant season to kill overwintering cleistothecia of grape powdery mildew. At dormant-season rates, it would be phytotoxic during the growing season. Eradicants such as triadimefon and myclobutanil have been developed that not only kill powdery mildew colonies but prevent sporulation as well. Many of these compounds also are active against rusts and various leaf-spotting fungi.

Many newer fungicides have systemic properties, which means they are absorbed and translocated by certain plant parts. Most of these fungicides are locally systemic. Green plant tissues such as leaves or shoots absorb the materials (such as DMI type of group 3 fungicides) and move them short distances within the transpiration stream (generally toward the leaf margin) or between plant cells. The QoI or strobilurin compounds (fungicide group 11) have a slightly different distribution. These compounds move into and through the leaf but do not move in the transpiration stream. This activity has been termed by the manufacturers as translaminar or quasi-systemic.

Long drying times and warmer temperatures after application favour the uptake of all of these materials.

Metalaxyl is a fungicide that can be absorbed by plant roots and translocated throughout the plant. Translocation, however, is only acropetally (upward), with the transpiration stream. Aliette is a truly systemic compound and is translocated both basipetally (downward) and acropetally whether applied to roots or leaves.

A fumigant is a vapour-active chemical used in the gaseous phase to kill microorganisms. Many nematicides are fumigants but kill fungal propagules as well as insects and weed seeds. Some fumigants such as sulfur are effective in the gaseous phase. Both temperature and humidity can influence the activity of sulfur in the vapour phase.

Antibiotics are produced by one organism against another. Many of the antibiotics in use today are bactericidal, but some also are fungicidal.

Some fungicides only inhibit fungi (are fungistatic) rather than kill them (fungicidal). Fungistats must be applied continually over the life of the plant to suppress disease development. For example, metalaxyl will prevent zoospores from penetrating roots but only inhibits an established *Phytophthora* infection.

Terms used to describe fungicides are not mutually exclusive. For example, dodine is an excellent broad-spectrum, protectant fungicide but also has 36-hour curative activity against apple scab.

Most fungicides directly inhibit or kill the target fungus, but compounds like Aliette can act by stimulating host defence mechanisms. Details of its activity in host resistance are poorly understood. These products are in general called plant activators.

Although disinfectants are considered good fungicides they are used primarily on inanimate surfaces or within irrigation lines. Some have over the plant uses. Unfortunately, they generally have short residual times and are not useful for long term fungal disease control.

Biomass Production of Spirulina

Spirulina is a microscopic blue-green alga in the shape of a spiral coil, living both in sea and fresh water. Spirulina is the common name for human and animal food produced primarily from two species of cyanobacteria: *Arthrospira platensis*, and *Arthrospira maxima*. Though referred to as 'algae' because they are aquatic organisms capable of photosynthesis, cyanobacteria are not related to any of the various eukaryotic algae.

Arthrospira is cultivated around the world, and is used as a human dietary supplement, as well as a whole food, and is available in tablet, flake, and powder form. It is also used as a feed supplement in the aquaculture, aquarium, and poultry industries.

Spirulina was a food source for the Aztecs and other Mesoamericans until the 16th century; its harvesting from Lake Texcoco and subsequent sale as cakes is described by one of Cortes' soldiers. The Aztecs called it Tecuitlatl, meaning stone's excrement.

Spirulina was found in abundance at Lake Texcoco by French researchers in the 1980s, but there is no reference to its use there as a daily food source after the 16th century. The first large-scale spirulina production plant, run by Sosa Texcoco, was established there in the early 1970s.

Leo Szilard postulated the development of algae-based food supplements (which he called "Amruss") in his 1961 short story, *The Voice of the Dolphins.*

Spirulina may have an even longer history in Chad, as far back as the 9th century Kanem Empire. It is still in daily use today, dried into cakes called *dihe*, which are used to make broths for meals, and also sold in markets. The spirulina is harvested from small lakes and ponds around Lake Chad.

Cultivation

Most cultivated spirulina is produced in open-channel raceway ponds, with paddle-wheels used to agitate the water. The largest commercial producers of spirulina are located in the United States, Thailand, India, Taiwan, China, Pakistan, Burma (a.k.a. Myanmar) and Chile.

Production of Algal Biomass

Algae (cyanobacteria and unicellular eukaryotes) grow autotrophically and synthesize their food by taking energy from sunlight or artificial light, carbon source from carbon dioxide, and nutrients from carbohydrates present in growth medium. In a few countries, cultivation of algae is carried-out in large trenches *i.e.* particularly in sewage oxidation ponds by using sunlight (Oswald, 1969) or in an artificial illumination conditions for use in life supportive systems for extended space exploration (Litchfield, 1967).

Chlorella strains are being used for a variety of applications in biotechnology. Due to their very high protein contents, they serve to improve protein deficiency and can be used as feed for production of animal protein. In many countries strains of *Chlorella* are utilized for sewage oxidation and waste water treatment (Kessler, 1989).

For cultivation of algae on sewage wastes, oxidation ponds are prepared, where sewage is allowed to accumulate. It is awaited till mixed cultures of algae grow (or inoculated with a singly prepared algal culture). For example, in Japan mixed culture of *Chlorella ellipsoides* and *Scenedesmus obliguus* was developed in open pond systems.

Factors Affecting Biomass Production

Following are the factors affecting the yield of biomass :

(i) Illumination time;

(ii) Light intensity;

(iii) Supply of CO_2. Concentrations of CO_2 differ in different conditions, for example, an alkaline lake. Lake Texcoco in Mexico, has high concentration of sodium carbonate. On the other hand, algal growth is limited as a result of liberation of CO_2 and ammonia by bacterial activity;

(iv) Nitrogen sources (ammonium salts or nitrates are the suitable nitrogen sources which increase biomass yield);

(v) Agitation of growing cells to maintain cells in suspension.

Biomass yield ranges between 12 and 15 g/m^2/day (on dry weight basis) as obtained with *Spirulina maxima* and *Scenedesmus quadricauda* grown in out door pond conditions (Clement, 1968).

Mass cultivation of algae has been started in many countries, such as Japan, West Germany (now Germany), Mexico, Chechoslovakia, India, etc. In India, National Environmental Engineering Research Institute (NEERI), Nagpur has developed a technique of cultivating algae in sewage oxidation pond systems. This practice is also in use at NBRI (Lucknow), Hyderabad and other centres. Interestingly, experiments conducted at the CFTRI, Mysore have shown that the microalgae *e.g. Scenedesmus acutus* and *Spirulina platensis* could be cultivated on a large scale and used as food and feed as they are rich in protein and their nutrient value is comparable to conventional foods.

Harvesting the Algal Biomass

Harvesting of algal culture becomes problematic because of settling down of cells at bottom and mixing of the algal cultures. The cells are recovered by concentration, dewatering and drying. Sometimes flocculants *e.g.* aluminium sulfate and calcium hydroxide and cationic polymers are added to the medium but they cannot be separated from the harvested cells. Therefore, this method warrants the application of SCP products in food and feed. Methods of separation and concentration also follow centrifugation, flocculation and centrifugation plus flocculation, but it is not economically feasible so far. Harvesting the cyanobacteria, for example, *Spirulina* sp. is less troublesome as their spiral filaments float on the surface of water because of gas filled vacuoles in their cells which result in floating algal mats. Cells are able to fix atmospheric nitrogen. Algal mats are filtered and suspension of *Spirulina* is dried with hot air to get fine powder. Algal yield from stabilization pond is around 114 tonnes/ha/year. In California 70 tonnes/ha/year of *Scenedesmus* was obtained from sewage. From Bankok much high yield of about 170 tonnes/ha/year has been reported (Vijayan, 1988).

SCP Product of Spirulina

In 1821, Bernal Diaz del Castillo, for described the biscuits with the name 'tecuitlatl'. It contained dried mats of *S. maxima* collected from the Lake Texcoco. The biscuit was sold in the markets of Mexico. In 1964, J. Leonard (a Belgian botanist) took part in an expedition and noticed dried biscuits of blue green colour in several village markets. The biscuit *dike* consumed by local population consisted of *Spirulina platensis.* Many pilot plants for the production of *Spirulina* powder have been established in Japan, United States and European countries. Sosa Texcoco is the first Mexican Company to set up the first pilot plant in 1973 which produced about 150 tonnes *Spirulina* powder per annum; but the yield was increased to 1,000 tonnes in 1982. This company exported powder to the United States and prepared lozenges and capsules from the powders by adding vitamin A and C. The Mexican Company also supplied *Spirulina* powder to government institutions which were in charge of improving the nutritional situation of the population. Institutions used *Spirulina* powder to make biscuits and confectionery with a high protein content (Sasson, 1984).

Benefits from Spirulina *SCP*

Mass cultivation of *Spirulatina* offers several advantages over *Chlorella* and *Scenedemus* as given below:

(*i*) Being a filamentous alga, *Spirulina* can be harvested by simple and less expensive methods such as nylon or cotton cloth filter.

(ii) Filaments of *Spirulina* float on water surface due to presence of gas vacuoles. Hence, there is no problem of harvesting unlike *Chlorella* and *Scenedesmus.*

(iii) There is least chance of contamination in growth tanks of *Spirulina* as it grows at high alkaline pH 8-11.

(iv) Heat drying is sufficient for *Spirulina* as it has thin cell wall, whereas spray drying is required for *Chlorella* and *Scenedesmus* which is expensive,

(v) Researches done by UNIDO programme in Mexico (1980), Mexican National Institute of Nutrition, Hyderabad (1988, 1990) under Indian Council of Medical Research on several aspects of possible adverse changes in multigeneration feeding tests on laboratory animals and humans have shown no adverse effects.

(vi) *Spirulina* is highly digestive (85-95%) due to thin wall and low nucleic acid contents (4%). It contains high percentage of digestible proteins (62-72%), vitamins, amino acids and other nutrients. The aminogram of *Spirulina* is comparable to the FAD, milk and egg protein pattern.

Table : Composition of multin* (i.e. *dried powder of* Spirulina fusiformis) *(constituents are in per 100 of powder)*.

A. Major constituents (%)		*C. Minerals (mg)*	
Total protein	64.6	Calcium	6.58
Fat	6.7	Phosphorus	977
Crude fibre ;	9.3	Iron	44.7
Carbohydrates	16.1	Sodium	796
Calories	346	Potassium	1.28
B. Vitamins		D. Essential amino acids (%)	
Beta- carotene	320,000IU	Lysin	2.99
Biotin	0.22 mg	Cystine	0.474
Cyanocobalamin (B_{12})	65.7 mg	Methionine	1.38
Folic acid	17.6 mg	Phenylalanine	2.87
Riboflavin	1.78 mg	Threonine	3.04
Thiamin	0.118 mg		
Tocopherol	0.773 IU		

* Analysed by Michelson Laboratories Inc., California, USA (1988)

Mass Cultivation of Spirulina SCP

At present two types of farms for mass production of *Spirulina* SCP are under operation. A third type *(i.e.* enclosed system using transparent tube, biocoil or photobioreacter) is under development (Henrikson, 1990).

(i) Semi-natural lake system. Sosa Texcoco Lake (Mexico) and Lake Chand (Africa) offer an ideal environment for the natural growth of *Spirulina.* The product is expensive but of low quality due to contamination and pollution by uncontrolled natural factors. SCP of these lakes are good for fish and animal feed. Researches are in progress to refine the powder and make the products of good quality.

(ii) Artificially built cultivation system. The climatic conditions of most of the developing countries are such that favour mass outdoor production of *Spirulina.* Therefore, on the basis of water quality and nutrient status, this system can be grouped into the following two :

(a) Clean water system. This system is more expensive because of construction of artificial cultivation farms. These have shallow raceway ponds circulated with paddle wheels and high quality of nutrients. For the fast growth of *Spirulina* in clear water, addition of $NaNO_3$ and $NaHCO_3$ is necessary. pH of the water must be initially maintained to 8.5. It is a self pH adjusting alga which elevates the pH between 10-10.5. At this pH levels there is the least chance of contamination.

The Earthrise Spirulina Farm of California is the world's largest food grade *Spirulina* farm having the size of about 10 hectares with a capacity of production to about 120 tonnes/annum. The other big farms are operating in Japan, Thailand, Mexico, Taiwan, Israel, Vietnam, China and India.

In India, food grade *Spirulina* is cultivated at two main centres, one at Shri A.M.M. Murugappa Chettiar Research Central (MCRC), Madras, and the other at Central Food Technology and Research Institute (CFTRI), Mysore. Madras centre is the biggest food grade *Spiulina* farm in India. Its annual production capacity is of about 75 tonnes. The products are marketed in India and abroad as health food, baby food and multivitamin tablets.

(b) Waste water system. This system is applicable in highly populated countries like India where wastes are generated in high quantities and pose environmental problem. In this system, human and animal wastes and sewage are used for growth of *Spirulina.* The wastes are added into the digester to settle down the solid particles. The liquid effluent is used as a source of the nutrients and added in artificially constructed ponds. As desired $NaNO_3$ and $NaHCO_3$ are also mixed. *S. platensis* is found to grow better in sewage amended with $NaCO_3$ and nutrients in different proportion and also in diluted sewage (Vijayan, 1988).

When full growth *of Spirulina* is over, it is screened from the pond and added to aquaculture to feed fish or dried in a small solar drier for human food.

This system is most suitable for third world countries where wastes are the major sources of pollution. R.D. Fox of France has developed the 'Integrated Health and Energy System Project' for poor countries to grow *Spirulina* and fight against the problems of food, malnutrition and environmental pollution. Today, a large number of these projects are in operation in the villages of developing countries like India, West Africa and South America. In India, the first integrated village system was established by Indo-France Govt. at Centre of Science for villages (CVS), Wardha (Maharastra).

The CVS distributes *Spirulina* cookies and nodules to malnourished children of local villages. This has shown very encouraging results.

Requirements for Growth of Spirulina

(i) Algal tanks. Generally, circular or rectangular cemented tanks are constructed. The circular tanks are more preferred over the rectangular one because of ease in handling. Size may be according to convenience and yield needed. Depth should be about 25cm. Open tanks are suitable for tropical and subtropical regions.

(ii) Light. Low light intensity is required at the beginning to avoid photolysis. *Spirulina* exposed to high light intensity is lysed.

(iii) Temperature. Temperature for optimum growth should be between 35-40°C.

(iv) pH. *Spirulina* grows at high pH ranging from 8.5 to 10.5. Initially, culture should be maintained at pH 8.5 which automatically is elevated to 10.5.

(v) Agitation. Agitation of culture is very necessary to get good quality and better yield. The culture is agitated by brush, paddle power, pipe pumps, wind power, rotators, etc.

(vi) Harvesting. The filaments of *Spirulina* float on surface of water forming thick mat. Therefore, it can be harvested by fine mesh steel screens, nylon or cotton cloths, etc.

(vii) Drying. As it has a thin wall, sun drying is the most suitable and economical. Various trials done at CFTRI, Mysore and MCRC, Madras with sun drying have given good results.

(viii) Yield. An average yield of 8-12 g *Spirulina* powder/m^2/day has been obtained in India and other countries. This is equivalent to 20 tonnes/ha/annum. In warmer climate, the yield can increased to about 20 g/m^2/day.

(ix) Avoiding contamination. Although there is the least chance for contamination, yet regular monitoring of algal culture is necessary. Because the microbial load is likely to affect the quality and safety of the product. At MCRC and CFTRI the cultures of *Spirulina* were found either within or very close to safety limits of Indian Standard Institute (ISI) for baby food, to about $5x10^5$ propagules per gram (Venkataraman and Becker, 1985). Dried *Spirulina* powder is packed in aluminium bags or sealed in bottles and sent to market.

Uses of *Spirulina* Single Cell Protein

(i) As protein supplemented food. Since *Spirulina* is a rich source of protein (60-72%), vitamins, amino acids, minerals, crude fibres, etc., it is used as supplemented food in diets of under-nourished poor children in developing countries (Jeegi Bai and Seshadri, 1986; Sachan, 1991). The United Nations, Mexican National Institute of Nutrition, French Petroleum Institute and National Institute of Nutrition, Hyderabad have formulated four algal recipes as a weaning substitute for infants. In India, the Village Health and Energy System Projects are operated at CVS (Wardha). At MCRC, the products are distributed to the local under-nourished children. It has been found that 1 g of *Spirulina* tablets contains as much nutrition as one kg assorted vegetable.

(ii) As health food. *Spirulina* is very popular as health food. Most of Sosa Texcoco products are exported to USA, Europe and France where it is sold in health and food stores. It is the part of the diet of the US Olympic team. Jaggers take *Spirulina* tablets for instant energy. Since it provides all the essential nutrients without excess calories and fats, it is taken by those who want to control obesity. The MCRC has for the first time launched the project as health and baby food, and multivitamin powder and tablet under trade name 'Multin' and 'Multinal'.

(iii) In the therapeutic and natural medicine. *Spirulina* possesses many medicinal properties. Therefore, it is used as social and preventive medicine also. It has been recommended by medicinal experts for reducing body weight, cholesterol and pre-menstrual stress and for better health. It lowers sugar level in blood of diabetics due to the presence of gamma-linolenic acid and prevents the accumulation of cholesterol in human body. It is a good source of (3-carotene (a precursor of vitamin A) and, therefore, helps in monitoring healthy eyes and skin. (3-carotene is known as the best anticancer substance. In 1989, UN National Cancer Research Institute announced that substances from blue-green algae are active against AIDS and cancer virus. In Vietnam its tablets are used to increase lactation in nourishing mothers.

(iv) In cosmetics. *Spirulina* contains high quality of proteins and vitamin A and B. These play a key role in maintaining healthy hair. Many herbal cosmeticians are making efforts to develop a variety of beauty products. Phycocyanin pigment has helped in formulating biolipstics and herbal face cream in Japan. These products can replace the present coaltar-dye based cosmetics which are known as carcinogenic.

Biological Role of Algae in Waste Water Treatment

Microorganisms form a very important part of the ecosystem. The diversity of microbial life found in the environment is quite amazing. Microbial life literally covers the planet. Microorganisms have a great influence on the biosphere. Chemosynthetic bacteria provide the energy and carbon for the survival of other organisms in regions where other organisms cannot reach. Microbes with the ability to recycle nutrients act as decomposers and have a special role in the biogeochemical cycles. Microbes are looked upon as decomposers of a variety of wastes. The ability to decompose is mainly dependent on the nature of the waste. The best way to approach microbial biodegradation is to use a mixture of bacterial species and strains due to the presence of different pollutant types at polluted sites. In order to evaluate the activity level it is important to monitor the composition of the indigenous and added bacteria.

Hydrocarbonoclastic bacteria (HCB) play a very important role in oil biodegradation. Their hydrocarbon degrading activities help to repair the damage caused by oil spills. They have applications in biotechnological areas such as bio plastics and biocatalysts. Bio treatment or the processing of waste by living organisms is an environmental friendly way for treating waste matter. Bioreactors have been designed to overcome the difficulties in bio treatment processes.

Wastewater treatment processes, too, use microbes for treatment. Microbes involved in nitrifying processes, anaerobic ammonia oxidation processes and methane fermenting processes are studies carefully for this purpose. The secondary treatment for wastewater involves the use of biological methods that approximate natural degradation processes. It sometimes includes chlorination to accomplish chemical oxidation and disinfection. In biological purification, the microbes bring about the following changes:

- Coagulation and flocculation of colloids and pseudocolloids
- Oxidation of carbonaceous matter to CO
- Nitrification i.e. oxidation of NH3 (derived from breakdown of nitrogenous organic matter to nitrite and then to nitrate).

The two main processes involved are trickling/percolation filters: these are manufactured beds of crushed stone used to percolate the waste water. They consist of circular or rectangular beds, about 1.8m deep of well-graded media. (clinker,stone,gravel,coke or even coal) usually about 40mm in size but increasing to 100-150mm for bottom enclosed in walls of brick or concrete. The filters bring the wastewater into contact with air and with the biological slime on the filter. The slime contains bacteria mainly zoogloea, fungi, protozoa, especially ciliates and other living organisms. The effective zone for aerobic bacterial activity is 2-3mm below the surface of the slime and their decomposition by the bacteria and fungi in the slime. Absorption and adsorption of solid and suspended matter takes place from the wastewater into and onto the slime. The slimes develop and proliferate on the surfaces of filtering medium over which the water trickles. This is followed by decomposition and synthesis of both the soluble substances and sludge forming solids. Near the top of the trickling filter, aerobic growth is important but further down, where less atmospheric oxygen diffuses, anaerobic action becomes predominant. When the slime layers become very thick, they slough off and can be removed. A large number of scouring animals like worms, flies and larvae are present in the filters which fed upon the biological film of slime preventing its overgrowth. The capacity of a filter can be increased by recirculation. This usually involves the biological filtration of settled sewage, which has been diluted, with a proportion of settled purified effluent from the filters.

Pumps are needed for Circulation

By recirculation, some microorganisms including enzymes with O2 and NO3 are then introduced into the influent waste. This refreshes the water, acts as a buffer against acidity, decreases odor and generally supports aerobic purification.

Activated Sludge Processes

This process consists essentially of aeration of a mixture of settled sewage with a special bacteriologically active sludge. The active sludge acts as a purifier. After settlement, the purified sewage is removed and some of the sludge is recirculated to assist the oxidation of a further quantity of sewage. As the process is repeated several times, the sludge becomes more active leading to the formation of a highly activated sludge, which can purify sewage in a few hours. The sludge performs two main functions:

- It clarifies the water by absorbing most of the colloidal suspended solids and surface of the sludge particles.
- It oxidizes the organic material.

It is important to ensure that the sludge in the mixture of sewage and activated sludge be kept in suspension by sufficient turbulence and that adequate amounts of oxygen are present. Oxidative degeneration is accomplished mostly by chemoheterotrophic bacteria and ciliated protozoa. Effective degradation is caused by genera like pseudomonas, zoogoea,flavorbacterium, alcallgenes etc.

The slime forming bacteria grow and form flocs. These flocs constitute activated sludge. They are composed of several kinds of bacteria mainly zoogloea ramigera, a small gram-negative rod that forms an extensive extracellular slime. This organism forms complex finger-like families that provide the basis of the floc. The other bacteria, protozoa etc attach to the floc and build up a complex active particle.

Biodisc System

The biodisc system is a more advanced type of aerobic film-flow treatment system. In this, closely spaced discs, usually made of plastic, are rotated in a trough containing the sewage effluent. The discs are particularly submerged and become coated with a slime similar to the one that develops in the trickling filters. Continuous rotation of the discs keeps the slime well aerated and in contact with the sewage. The thickness of the slime layer is related to the diffusion of nutrients through the film. Microbial growth on the surface of the disc is sloughed off gradually and is removed by subsequent settling. When the film becomes so thick that O2 nutrients fail to reach the inner portions of the film, most of the innermost microorganisms die causing detachment to the slime layer. Patches of slime periodically fall off the surface. However, the slime layer begins to grow immediately.

Biodisc systems can be used to treat both domestic and industrial sewage effluents. It requires less space than trickling filters and is more efficient and stable in operation. However, it requires higher financial investment initially.

Oxidation Ponds

Also known as stabilization ponds, or lagoons, these are used for the simple secondary treatment of sewage effluents in rural communities and some industries. Such ponds are inexpensive to build and operate but require large areas of land. Generally, a 2 stage operation ensues. The 1st stage is analogous to primary treatment; the sewage pond is deep enough so that conditions are anaerobic. Sludge settles down in this stage. In the 2nd stage, which roughly corresponds to secondary treatment; effluent is pumped into an adjoining pond/s shallow enough to be aerated by wave action.

Heterotrophic bacteria (organisms requiring organic compounds for growth and reproduction, the organic compounds serve as sources of carbon and energy) degrade sewage organic matter within the ponds, producing cellular material and mineral products that support the growth of algae. The proliferation of algal population in these lagoons produces oxygen that replenishes the oxygen depleted by the heterotrophic bacteria, permitting continued organic matter decomposition. Because oxygenation is usually achieved by diffusion and by the photosynthetic activity of algae, such ponds need to be shallow. Typically, these are less than 10 feet deep which maximizes the euphoric zone for algal growth. The degradation of organic matter in these ponds is relatively slow and residence time for the treatment of domestic sewage maybe as long as a week. If properly operated the ponds are reasonably free from bad smells, possibly due to deodorizing effect of the chlorophyll in the algae. Treatment of sewage in such ponds usually gives a good reduction in BOD. Algae from these ponds are rich in proteins and fat may be used as cheap animal feed.

Septic Tanks

This is the simplest anaerobic treatment system, used extensively in rural areas that lack sewage systems. A septic tank acts largely as a settling tank within which the organic components of the wastewater undergo limited anaerobic digestion, being converted to organic acids and H2S. Thus, it is a single –storey-settling tank, where settled sludge is in immediate contact with sewage flowing through the tank while solids are being decomposed by anaerobic bacterial action. Residual solids settle to the bottom of the septic tank and clarified effluent flow through a series of perforated piping into a leaching (soil drainage) field. Soil microorganisms decompose the effluent entering the soil.

These systems work well when not overloaded. Septic tank treatment does not reliably destroy intestinal pathogens and it is important that the soil receiving the clarified effluents is not in close proximity to drinking wells to prevent contamination. Efficiency in removal of solids is less than that for plain sedimentation. While 60% TSS should be removed theoretically, it is seldom achieved in practice. These are generally designed to hold 12-24 hr flow.

Examples of Microbiological Utilities of Solid Wastes

Biogas Production from Pineapple Wastes in Anaerobic Conditions

Approximately 50% of raw material is waste. Juice is squeezed from peel by using screw press, filtered and transferred to filter reactors. It is left there for thirty days to allow the colonies of bacteria to be formed. CH4 is formed from the utilization of glucose etc by the microorganisms.

Use of Coconut Oil Meal as Animal Feed

Coconut oil meal is the ground residue left after the extraction of oil from the dried coconuts. It is rich in proteins and carbohydrates and can be used as animal feed. However, it has high fibre content, which decreases the digestibility of the meal. This can be removed by treatment the copra meal with mannanase from Streptomyces.

Use of Agricultural Residues for Mushroom Cultivation

Agricultural wastes like soybean husk, mungbean husk, and coffee husk as also other agricultural and household solid wastes rich in cellulose, hemi cellulose and lignin can be used as growth substrate for mushroom (Agaricus bisporus, Pleurotus sps)after proper treatment with microorganisms.

Use of Waste Soybean Slurry to Produce Mycomeat

Waste soybean slurry (high in protein) can be used to form an edible product called mycomeat, after treatment with certain fungi. The soy slurry is first squeezed through cheesecloth or Morchelle crassipes or remove water. It is then placed in a culture dish of depth of less than 1 inch and inoculated with strains (blocks of mycelium grown on agar). Incubation is at 25°C for varying lengths of time. The mycomeat obtained has a good taste, texture and high amounts of proteins (27%), carbohydrates (31%) and lipids (7%).

Citric Acid Produced from Fruit Pomace

Fruit pomace is a byproduct of the fruit juice processing industry. It is rich in carbohydrates. Some strains of Apergillus niger produce large amounts of citric acid when allowed to grow on apple and grape pomace in static flask cultures.

5

Biogas Technology

Biogas typically refers to a gas produced by the biological breakdown of organic matter in the absence of oxygen. Biogas originates from biogenic material and is a type of biofuel. Biogas is produced by anaerobic digestion or fermentation of biodegradable materials such as biomass, manure, sewage, municipal waste, green waste, plant material and energy crops. This type of biogas comprises primarily methane and carbon dioxide. Other types of gas generated by use of biomass is wood gas, which is created by gasification of wood or other biomass. This type of gas consist primarily of nitrogen, hydrogen, and carbon monoxide, with trace amounts of methane. The gases methane, hydrogen and carbon monoxide can be combusted or oxidized with oxygen. Air contains 21% oxygen. This energy release allows biogas to be used as a fuel. Biogas can be used as a low-cost fuel in any country for any heating purpose, such as cooking. It can also be used in modern waste management facilities where it can be used to run any type of heat engine, to generate either mechanical or electrical power. Biogas can be compressed, much like natural gas, and used to power motor vehicles and in the UK for example is estimated to have the potential to replace around 17% of vehicle fuel. Biogas is a renewable fuel, so it qualifies for renewable energy subsidies in some parts of the world.

Ancient Persians observed that rotting vegetables produce flammable gas. In the 13th century, the traveller Marco Polo noted the Chinese used covered sewage tanks to generate power, while biogas technologies were also referred to by 17th century author Daniel Defoe.

In 1859, an anaerobic digestion plant was built to process sewage at a Bombay leper colony. Biogas has been used in the UK since 1895, when gas from sewage was used in street lamps across the city of Exeter.

Production

Biogas is practically produced as landfill gas (LFG) or digester gas.

A biogas plant is the name often given to an anaerobic digester that treats farm wastes or energy crops.

Biogas can be produced utilizing anaerobic digesters. These plants can be fed with energy crops such as maize silage or biodegradable wastes including sewage sludge and food waste. During the process, an air-tight tank transforms biomass waste into methane producing renewable energy that can be used for heating, electricity, and many other operations that use any variation of an internal combustion engine, such as GE Jenbacher gas engines. There are two key processes: Mesophilic and Thermophilic digestion. Landfill gas is produced by wet organic waste decomposing under anaerobic conditions in a landfill. The waste is covered and mechanically compressed by the weight of the material that is deposited from above. This material prevents oxygen exposure thus allowing anaerobic microbes to thrive. This gas builds up and is slowly released into the atmosphere if the landfill site has not been engineered to capture the gas. Landfill gas is hazardous for three key reasons. Landfill gas becomes explosive when it escapes from the landfill and mixes with oxygen. The lower explosive limit is 5% methane and the upper explosive limit is 15% methane. The methane contained within biogas is 20 times more potent as a greenhouse gas than carbon dioxide. Therefore uncontained landfill gas which escapes into the atmosphere may significantly contribute to the effects of global warming. In addition landfill gas' impact in global warming, volatile organic compounds (VOCs) contained within landfill gas contribute to the formation of photochemical smog.

Composition

Typical composition of biogas

Compound	*Chem*	*%*
Methane	CH_4	50–75
Carbon dioxide	CO_2	25–50
Nitrogen	N_2	0–10
Hydrogen	H_2	0–1
Hydrogen sulfide	H_2S	0–3
Oxygen	O_0	0–0

The composition of biogas varies depending upon the origin of the anaerobic digestion process. Landfill gas typically has methane concentrations around 50%. Advanced waste treatment technologies can produce biogas with 55–75% CH_4 or higher using in situ purification techniques As-produced, biogas also contains water vapour, with the fractional water vapour volume a function of biogas temperature; correction of measured volume for water vapour content and thermal expansion is easily done via algorithm.

In some cases biogas contains siloxanes. These siloxanes are formed from the anaerobic decomposition of materials commonly found in soaps and detergents. During combustion of biogas containing siloxanes, silicon is released and can combine with free oxygen or various other elements in the combustion gas. Deposits are formed containing mostly silica (SiO_2) or silicates (Si_xO_y) and can also contain calcium, sulfur, zinc, phosphorus. Such white mineral deposits accumulate to a surface thickness of several millimeters and must be removed by chemical or mechanical means.

Practical and cost-effective technologies to remove siloxanes and other biogas contaminants are currently available.

Applications

Biogas can be utilized for electricity production on sewage works, in a CHP gas engine, where the waste heat from the engine is conveniently used for heating the digester; cooking; space heating; water heating; and process heating. If compressed, it can replace compressed natural gas for use in vehicles, where it can fuel an internal combustion engine or fuel cells and is a much more effective displacer of carbon dioxide than the normal use in on-site CHP plants.

Methane within biogas can be concentrated via a biogas upgrader to the same standards as fossil natural gas(which itself has had to go through a cleaning process), and becomes biomethane. If the local gas network allows for this, the producer of the biogas may utilize the local gas distribution networks. Gas must be very clean to reach pipeline quality, and must be of the correct composition for the local distribution network to accept. Carbon dioxide, water, hydrogen sulfide and particulates must be removed if present. If concentrated and compressed it can also be used in vehicle transportation. Compressed biogas is becoming widely used in Sweden, Switzerland, and Germany. A biogas-powered train has been in service in Sweden since 2005.

Biogas has also powered automobiles. In 1974, a British documentary film entitled *Sweet as a Nut* detailed the biogas production process from pig manure, and how the biogas fueled a custom-adapted combustion engine.

Benefits

By using biogas, many advantages arise. In North America, utilization of biogas would generate enough electricity to meet up to three percent of the continent's electricity expenditure. In addition, biogas could potentially help reduce global climate change. Normally, manure that is left to decompose releases two main gases that cause global climate change: nitrous dioxide and methane. Nitrous dioxide warms the atmosphere 310 times more than carbon dioxide and methane 21 times more than carbon dioxide. By converting cow manure into methane biogas via anaerobic digestion, the millions of cows in the United States would be able to produce one hundred billion kilowatt hours of electricity, enough to power millions of homes across the United States. In fact, one cow can produce enough manure in one day to generate three kilowatt hours of electricity; only 2.4 kilowatt hours of electricity are needed to power a single one hundred watt light bulb for one day. Furthermore, by converting cow manure into methane biogas instead of letting it decompose, we would be able to reduce global warming gases by ninety-nine million metric tons or four percent.

The 30 million rural households in China that have biogas digesters enjoy 12 benefits: saving fossil fuels, saving time collecting firewood, protecting forests, using crop residues for animal fodder instead of fuel, saving money, saving cooking time, improving hygienic conditions, producing high-quality fertilizer, enabling local mechanization and electricity production, improving the rural standard of living, and reducing air and water pollution.

Biogas Upgrading

Raw biogas produced from digestion is roughly 60% methane and 29% CO_2 with trace elements of H_2S, and is not high quality enough if the owner was planning on selling this gas or using it as fuel gas for machinery. The corrosive nature of H_2S alone is enough to destroy the internals of an expensive plant. The solution is the use of a biogas upgrading or purification process whereby contaminants in the raw biogas stream are absorbed or scrubbed, leaving 98% methane per unit volume of gas. There are four main methods of biogas upgrading, these include water washing, pressure swing absorption, selexol absorption and chemical treatment. The most prevalent method is water washing where high pressure gas flows into a column where the carbon dioxide and other trace elements are scrubbed by cascading water running counter-flow to the gas. This arrangement can deliver 98% methane with manufacturers guaranteeing maximum 2% methane loss in the system. It takes roughly between 3-6% of the total energy output in gas to run a biogas upgrading system.

Biogas Gas-grid Injection

Gas-grid injection is the injection of biogas into the methane grid (natural gas grid). Injections includes biogas: until the breakthrough of micro combined heat and power two-thirds of all the energy produced by biogas power plants was lost (the heat), using the grid to transport the gas tc customers, the electricity and the heat can be used for on-site generation resulting in a reduction of losses in the transportation of energy. Typical energy losses in natural gas transmission systems range from 1–2%. The current energy losses on a large electrical system range from 5–8%.

In the field of renewable energy, biogas technology refers to systems designed to turn organic waste products into usable energy. Biogas is a kind of gas that is produced during the anaerobic processing of organic matter such as manure, plant matter, or even municipal waste materials. Biogas typically consists mainly of methane, with a significant proportion of carbon dioxide, and smaller quantities of other gases such as nitrogen and hydrogen.

Biogas fuel is a flammable substance that burns in a similar fashion to liquefied petroleum gas (LPG), and as such, biogas energy can be utilized as an alternative to fossil fuels. Biogas production is often achieved using a biogas plant, which is a system that "digests" organic matter to produce gas. Biogas technology is often used on farms with the primary goal of controlling waste pollution. Dairy farms, for example, may have significant problems with manure polluting water sources. Biogas technology allows farmers to address this problem with the beneficial side effect of also creating a new power source.

Biogas electricity and energy for heating can be produced locally with biogas technology systems on some farms. For example, a farmer with 500 cows could install a system that would process the manure produced by the cows, and could generate sufficient electricity to power the entire dairy system. Larger scale systems in some countries collect manure from multiple farms, and perform the digestion process centrally in a large plant.

Similar biogas technology may be used in sewage treatment. Wastewater contains organic matter, including organic solids, and anaerobic digestion may be used to break down these solids. This process, often called sludge digestion, may also reduce the levels of harmful bacteria in the water.

There are three main types of biogas technology digesters: covered lagoon, complete mix, and plug-flow. All of these types digest manure to produce biogas. They differ in their efficiency, the concentration of solids they can handle, and their suitability to different climates and applications.

Covered lagoon digesters are the simplest type, consisting of a storage pool to hold the manure, and a cover to trap the gas. This type of technology is suitable for warm climates, and is mainly used for liquid manure with a small percentage of solid matter. Complete mix digesters hold manure in a heated tank, and have a mechanical mixing device to speed up the digestion process. Plug-flow digesters are more complex biogas technology systems that actively pass the manure being processed through the system, as it is digested.

- Dairy farms, for example, may have significant problems with manure polluting water sources. Biogas technology allows farmers to address this problem with the beneficial side effect of also creating a new power source.
- Unintentional production of biogases has been an ongoing issue in many regions of the world, and several nations have also looked to biogas as a potential source of clean energy. India and China have both invested extensively in creative biogas technology to provide fuel for their citizens and there are a number of interesting applications for this gas which appeal to people who are interested in sustainable energy and the health of the environment.

Biogas Plants for Organic Farming

The biogas plant consists of two components: a digester (or fermentation tank) and a gas holder. The digester is a cube-shaped or cylindrical waterproof container with an inlet into which the fermentable mixture is introduced in the form of a liquid slurry. The gas holder is normally an airproof steel container that, by floating like a ball on the fermentation mix, cuts off air to the digester (anaerobiosis) and collects the gas generated. In one of the most widely used designs, the gas holder is equipped with a gas outlet, while the digester is provided with an overflow pipe to lead the sludge out into a drainage pit.

Singh, and others (1, 3) have documented several guidelines for consideration in the designing of batch (periodic feeding) and continuous (daily feeding) compartmentalized and non-compartmentalized biogas plants that are of either the vertical or horizontal type. In addition, Loll (18) has recently dealt with the scientific principles, process engineering, and shapes of digestion reactors, and with the economics of the technology.

Digester reactors are constucted from brick, cement, concrete, and steel. In Indonesia, where rural skills in brick making, brick laying, plastering, and bamboo craft are well established, clay bricks have successfully replaced cement blocks and concrete. In areas where the cost is high, the "sausage" or bag digester (14) appears to be ideal. The digester is constructed of 0.55 mm thick Hypalon laminated with Neoprene and reinforced with nylon. The bag is fitted with an inlet and an outlet made from PVC. Even if imported from the United States, the cost of the digester and the gas holder (both combined in one bag) is only 10 per cent of that for a concrete-steel digester. Another advantage is that it can be mass produced and is easily mailed. In rural areas, the whole installation is completed in a matter of minutes. A hole in the ground accommodates the bag, which is filled two-thirds full with waste water. Gas production fully inflates the bag, which is weighted down and fitted with a compressor to increase gas pressure.

Environmental and Operational Considerations

Raw Materials (19)

Raw materials may be obtained from a variety of sources- livestock and poultry wastes, night soil, crop residues, food-processing and paper wastes, and materials such as aquatic weeds, water hyacinth, filamentous algae, and seaweed. Different problems are encountered with each of these wastes with regard to collection, transportation, processing, storage, residue utilization, and ultimate use. Residues from the agricultural sector such as spent straw, hay, cane trash, corn and plant stubble, and bagasse need to be shredded in order to facilitate their flow into the digester reactor as well as to increase the efficiency of bacterial action. Succulent plant material yields more gas than dried matter does, and hence materials like brush and weeds need semi-drying. The storage of raw materials in a damp, confined space for over ten days initiates anaerobic bacterial action that, though causing some gas loss, reduces the time for the digester to become operational.

Influent Solids Content (16, 19, 21)

Production of biogas is inefficient if fermentation materials are too dilute or too concentrated, resulting in, low biogas production and insufficient fermentation activity, respectively. Experience has shown that the raw-material (domestic and poultry wastes and manure) ratio to water should be 1:1, i.e., 100 kg of excrete to 100 kg of water. In the slurry, this corresponds to a total solids concentration of 8- 11 per cent by weight.

Loading (14, 19)

The size of the digester depends upon the loading, which is determined by the influent solids content, retention time, and the digester temperature. Optimum loading rates vary with different digesters and their sites of location. Higher loading rates have been used when the ambient temperature is high. In general, the literature is filled with a variety of conflicting loading rates. In practice, the loading rate should be an expression of either (a) the weight of total volatile solids (TVS) added per day per unit volume of the digester, or (b) the weight of TVS added per day per unit weight of TVS in the digester. The latter principle is normally used for smooth operation of the digester.

Seeding (14, 19)

Common practice involves seeding with an adequate population of both the acid-forming and methanogenic bacteria. Actively digesting sludge from a sewage plant constitutes ideal "seed" material. As a general guideline, the seed material should be twice the volume of the fresh manure slurry during the start-up phase, with a gradual decrease in amount added over a three-week period. If the digester accumulates volatile acids as a result of overloading, the situation can be remedied by reseeding, or by the addition of lime or other alkali.

pH (14, 19)

Low pH inhibits the growth of the methanogenic bacteria and gas generation and is often the result of overloading. A successful pH range for anaerobic digestion is 6.0- 8.0; efficient digestion occurs at a pH near neutrality. A slightly alkaline state is an indication that pH fluctuations are not too drastic. Low pH may be remedied by dilution or by the addition of lime.

Temperature (13,14,19, 21)

With a mesophilic flora, digestion proceeds best at 30- 40 C; with thermophiles, the optimum range is 50- 60 C. The choice of the temperature to be used is influenced by climatic considerations In general, there is no rule of thumb, but for optimum process stability, the temperature should be carefully regulated within a narrow range of the operating temperature. In warm climates, with no freezing temperatures, digesters may be operated without added heat.

As a safety measure, it is common practice either to bury the digesters in the ground on account of the advantageous insulating properties of the soil, or to use a greenhouse covering. Heating requirements and, consequently, costs, can be minimized through the use of natural materials such as leaves, sawdust, straw, etc., which are composted in batches in a separate compartment around the digester,

Nutrients (13,17,19, 21)

The maintenance of optimum microbiological activity in the digester is crucial to gas generation and consequently is related to nutrient availability. Two of the most important nutrients are carbon and nitrogen and a critical factor for raw material choice is the overall C/N ratio.

Domestic sewage and animal and poultry wastes are examples of N-rich materials that provide nutrients for the growth and multiplication of the anaerobic organisms. On the other hand, N-poor materials like green grass, corn stubble, etc., are rich in carbohydrate substances that are essential for gas production. Excess availability of nitrogen leads to the formation of NH3, the concentration of which inhibits further growth. Ammonia toxicity can be remedied by low loading or by dilution. In practice, it is important to maintain, by weight, a C/N ratio close to 30:1 for achieving an optimum rate of digestion. The C/N ratio can be judiciously manipulated by combining materials low in carbon with those that are high in nitrogen, and vice versa.

Toxic Materials (13,14,19)

Wastes and biodegradable residue are often accompanied by a variety of pollutants that could inhibit anaerobic digestion. Potential toxicity due to ammonia can be corrected by remedying the C/N ratio of manure through the addition of shredded bagasse or straw, or by dilution. Common toxic substances are the soluble salts of copper, zinc, nickel, mercury, and chromium. On the other hand, salts of sodium, potassium, calcium, and magnesium may be stimulatory or toxic in action, both manifestations being associated with the cation rather than the anionic portion of the salt. Pesticides and synthetic detergents may also be troublesome to the process.

Stirring (13,14,17- 19, 21)

When solid materials not well shredded are present in the digester, gas generation may be impeded by the formation of a scum that is comprised of these low-density solids that are enmeshed in a filamentous matrix. In time the scum hardens, disrupting the digestion process and causing stratification. Agitation can be done either mechanically with a plunger or by means of rotational spraying of fresh influent. Agitation, normally required for bath digesters, ensures exposure of new surfaces to bacterial action, prevents viscid stratification and slow-down of bacterial activity, and promotes uniform dispersion of the influent materials throughout the fermentation liquor, thereby accelerating digestion.

Retention Time (19, 21)

Other factors such as temperature, dilution, loading rate, etc., influence retention time. At high temperature bio-digestion occurs faster, reducing the time requirement. A normal period for the digestion of dung would be two to four weeks.

Developments and Processes for Rural Areas

Two years ago, the Economic and Social Council of the United Nations adopted a survey, presented in 1978 to the Committee on Science and Technology for Development, listing the on-going research and development in unconventional sources of energy. From the point of view of the developing countries, it is heartening to note that the "use of farm wastes to produce methane" has also been identified in the United Nations World Plan of Action for the Application of Science and Technology to Development.

The Economic and Social Council for Asia and the Pacific, moreover, adopted the Colombo Declaration at its thirtieth session, which determined that the most urgent priorities for action are in the fields of food, energy, raw materials, and fertilizers, and that these priorities would be best met by the integrated biogas system (IBS).

An integrated system aims at the facile generation of fertilizer and acquisition of energy, production of protein via the growth of algae and fish in oxidation ponds, hygienic disposal of sewage and other refuse, and is a tangible effort to counteract environmental pollution. The heart of the system is the biogas process; it has the potential to "seed" self-reliance in relatively primitive economies (14, 22, 23). Allied benefits include the development of rural industry, the provision of local job opportunities, and the progressive eradication of hunger and poverty.

The coupling of a photosynthetic step (24- 26) with digestion provides for the transformation of the minerals left by digestion directly into algae that can then be used as fodder, as feed for fish, as fertilizer, or for increased energy production by returning them to the digester process.

The IBS aims at putting back into soil and water what has been taken from them, and increasing the amounts of nutrients by fixing CO2 and N2 from the atmosphere into the soil and water through photosynthesis by algae. Involving low cash investments on a decentralized basis, the implementation of IBS provides employment to the whole work force without disruption of the rural structure. Furthermore, it is an apt example of soft technology that does not pollute or destroy the physical environment. At the College of Agriculture of the University of the Philippines, preliminary work on a small scale has begun. In England, an Eco-house has been built by Graham Caine on the Thames Polytechnical Playing Fields at Eltham, southeast of London. Results on the project, however, are not yet available.

Cost-benefit Analyses

There is no general answer to the economic feasibility of biogas production. National economic considerations play an important role. In Korea, wood is in short supply (27) and domestic fuel substitutes like rice and barley straw, and coal and oil could be conserved; wood could be a foreign-exchange earner in the field of handicrafts. In India, transportation costs of coal and oil to the rural areas is high and an extra burden on an already poor farmer.

The rural share in the energy consumption of electricity and coal is not considerable because, as the Report of the Panel of the National Committee of Science and Technology on Fuel and Power indicates, the large towns and cities with populations of 500,000 and more accommodate only 6 per cent of India's total population but consume about 50 per cent of the total commercial energy produced in the country.

In the villages, however, kerosene is used for lighting, but it is clear that with increasing population, biogas generation seems to offer solutions in the areas of fuel availability, electricity, fertilizer for cash crops, and would provide other socio-economic benefits.

On the other hand, cost-benefit analyses of methane generation vary widely, depending upon the uses and actual benefits of biogas production, public and private costs associated with the development and utilization of methane, and on the technology used to generate methane. Several factors have been listed in the economics of biogas generation (14, 17- 19, 28). An appropriate example is the fact that a village-model gas plant, which cost Rs 500 some years ago, cost Rs 1,500 in 1974 and Rs 2,000 in 1977. Hence, a significant problem is whether rural people who cannot spend Rs 2,000 can cope with increasing inflationary and digester construction material costs.

The Khadi and Village Industries Commission has helped to tackle the problem through rural community co-operation and a scheme of subsidies and loans to encourage individual families, groups of families, institutions, and communities to construct biogas plants. The net annual income of approximately US$60 shows that the capital investment of US$340 can be recouped in about six years. There are also incidental advantages of hygienic improvement, the absence of smoke and soot in gas burning, convenience in burning, and the increased richness of manure.

Health Hazards

Health hazards are associated with the handling of night soil and with the use of sludge from untreated human excrete as fertilizer. In general, published data indicate that a digestion time of 14 days at 35 C is effective in killing (99.9 per cent die-off rate) the enteric bacterial pathogens and the enteric group of viruses.

However, the die-off rate for roundworm (Ascaris lumbricoides) and hookworm (Ancylostoma) is only 90 per cent, which is still high. In this context, biogas production would provide a public health benefit beyond that of any other treatment in managing the rural health environment of developing countries.

Production Technology of Paddy Straw Mushroom

The steps involved in cultivation are; choice of substratum, bed-preparation and cropping, after care of beds, and harvesting and marketing

Choice of Substratum

Paddy straw is considered to be the best; other which can be used are wheat straw, jowar, maize, rye straw, sugarcane by gasse, tobacco, and banana leaves

Bed-preparation and Cropping

Firstly, hand-harvested, 3-4' long. Well dried and disease-free paddy straw are taken to prepare their bundles. 35 bundles of paddy straw are required for one bed and each bundle should be 1 to 11/2 kg in weight.

The bundles are soaked in water for 8-16 hours, are taken out of water, washed with fresh water and allowed to drain off excess water.

Now, a bed is prepared by putting four layers of the paddy straw bundles one over the other each layer contains 8 bundles. The spawn bottle is opened and raked with a woolen or glass rod.

The spawn is now sprinkled by hand all over on the margin of the bed about 10 cms. Away from the edge and continuing upto 23 cms. Inside. Thus, the central portion has been left spawning in 1st layer. The sprinkled spawn is covered with a light dusting of 'besan' (gram powder). The second, third and fourth layers are also prepared in the same way as in case of 1st but the difference in 4th layer is that the sprinkling of spawn is done on the entire surface instead of periphery as in case of rests of the layers. The layer should also be dusted by the gram powder and must be covered with a thin layer of straw upto 8 cms. Thickness. This is also done with the remaining three bundles of straw.

After Care of Beds

Water is sprayed 2-3 times in hot day and 1-2 times in rainy season. If necessary, 0.1% Malathion and 0.2% Dithane Z-78 is sprayed to overcome insects, pests, and other diseases.

Harvesting and Marketing

Cropping starts 10-12 days after spawning and remains upto 15-20 days. Mushroom is harvested at button stage or just after rupturing of the cup (pileus). The harvesting is done by twisting so that broken pieces are not left in the beds otherwise bacterial rotting generally starts and spreads to other healthy mushrooms.

After harvesting the mushrooms should be used within 8 hours or kept 10-15^{0}C for 24 hours otherwise they get spoiled. One may keep them for a week in refrigerator.

Fresh mushrooms are dried either in sun or in even at 55-60^{0}C for 8 hours. After drying, they are packed and sealed otherwise they absorb moisture which spoils them. The packing of mushrooms is preferred only in button stage. However, one can find 3-4 kg yield/bed of mushrooms.

Microbial Process in Mushroom Production

The mushroom is a fungus and is quite finicky about its food source. Mushrooms lack the ability to use energy from the sun. They are not green plants because they do not have chlorophyll. Mushrooms extract their carbohydrates and proteins from a rich medium of decaying, organicmatter vegetation. This rich organic matter must be prepared into nutrient- rich substrate composts that the mushroom can consume. When correctly made, this food may become available exclusively to the mushroom and would not support the growth of much else. At a certain stage in the decomposition, the mushroom grower stops the process and plants the mushroom so it becomes the dominant organism in that environment.

The sequence used to produce this specific substrate for the mushroom is called composting or compost substrate preparation and is divided into two stages, Phase I and Phase II. Each stage has distinct goals or objectives. It is the grower's responsibility to provide the necessary ingredients and environmental conditions for the chemical and biological processes required to complete these goals. The management of starting ingredients and the proper conditions for composting make growing mushrooms so demanding.

Making a Composted Substrate

Many agricultural by-products are used to make mushroom substrate. Straw-bedded horse manure and hay or wheat straw are the common bulk ingredients. "Synthetic" composts are those in which the prime ingredient is not straw-bedded horse manure. If bulk ingredients are high in nitrogen, other high-carbohydrate bulk ingredients—such as corncobs, cottonseed hulls, or cocoa bean hulls—are added to the mix. All compost formulas require the addition of nitrogen supplements and gypsum. Additional nitrogen-rich supplements are added to composts to increase the nitrogen content to 1.5–1.7 percent for horse manure or 1.7–1.9 percent for synthetic; both are computed on a dry weight basis. Poultry manure is probably the most common and economical source of nitrogen. A variety of meals or seeds, such as cottonseed meal, soybean meal, or brewer's grain may also be used. Inorganic or nonprotein nitrogen sources such as ammonia nitrate and urea are also used, but only in small amounts when high-carbohydrate bulk ingredients are used. Gypsum is added to minimize "greasiness" and to buffer the pH of the compost.

Gypsum increases the flocculation of colloids in the compost, which prevents the straws from sticking together and inhibiting air penetration.

Air, which supplies oxygen to the microbes and chemical reactions, is essential to the composting process. Gypsum may be added early in the composting process, at 70–100 lbs per ton of dry ingredients.

A concrete slab, referred to as a wharf, is required for composting. In addition, a compost turner to aerate and water the ingredients and a tractor-loader to move the ingredients to the turner are needed. Water used during a substrate preparation operation can be recycled back into the process. It is, in a sense, a closed system. Water runoff into the environment is nonexistent on a properly managed substrate preparation wharf. Water collected in concrete pits or a sealed lagoon is aerated and recycled to soak bulk ingredients before the composting process begins.

Conventional Phase I composting begins by mixing and wetting the ingredients as they are stacked. Most farms have a preconditioning phase in which bulk ingredients and some supplements are watered and stacked in a large pile for several days to soften, making them more receptive to water. This preconditioning time may range from 3 to 15 days. The piles are turned daily or every other day. After this pre-wet stage, the compost is formed into a rectangular pile with tight sides and a loose centre.

A compost turner is typically used to form this pile. Water is sprayed onto the horse manure or synthetic compost as these materials move through the turner. Nitrogen supplements and gypsum can be spread over the top of the bulk ingredients and are thoroughly mixed by the turner.

Once the pile is wetted and formed, aerobic fermentation (composting) commences as microbial growth and reproduction naturally occur in the bulk ngredients. Heat, ammonia, and carbon dioxide (CO2) are released as by-products during this process. Compost activators, other than those mentioned, are not needed.

As temperatures increase above 155°F (70°C), microorganisms cease growing and a chemical reaction begins.

Concentrating and preserving complex carbohydrates is one goal of Phase I. The quantity and the quality of nitrogen in the system are changed to a type of nitrogen that Phase II microorganisms and, eventually, the mushroom will use as food.

Adequate moisture, oxygen, nitrogen, and carbohydrates must be present throughout the process; otherwise, the process will stop. This is why water and supplements are added periodically and the compost pile is aerated as it moves through the turner. Oxygenation is achieved in conventional outdoor ricks by natural convection.

The high pile temperatures draw ambient air through the sides of the stack, and as the air is heated, it rises upward through the stack—a process commonly referred to as the chimney effect. The sides of the pile should be firm and dense, yet the centre must remain loose throughout Phase I composting. The exclusion of air results in an airless (anaerobic) environment. As the straw or hay softens during composting, the materials become less rigid and more compact while substrate density increases. Thus, less air reaches the bottom and centre of the pile. A lack of oxygen may occur after the large quantities of water are added to the dry bulk ingredients and before sufficient heat is generated to start the draw of air into the pile. Under anaerobic conditions, organic acids and other deleterious chemical compounds are formed. Therefore, preparing substrate under aerobic conditions, where less offensive odors are produced, is better for mushroom growers.

Aerated Phase I Composting

Improving community relations has led to alterations in the way the Phase I mushroom composting process is carried out. As urban areas encroach on rural farmland, residents have made odor-related complaints and legal battles have ensued, which suggest a need for more stringent odor-management practices.

If the pile is not turned and aerated during Phase I composting, oxygen may become limited and anaerobic conditions may develop along the bottom of the stack. As the anaerobic core gets larger, more offensive odors are produced. In order to maintain aerobic conditions throughout the entire substrate pile, supplemental aeration is sometimes used. This aeration is accomplished by using a fan to force air up through a concrete pad with a series of evenly distributed openings and into the substrate material. This design is referred to as an aerated floor. Systems have been built with structural sidewalls, usually of concrete and occasionally of wood, to form the piles with a uniform height and depth. Aside from aerated floors and structural sidewalls, there is great variation among bunker systems currently being used for Phase I.

Aerated composting systems are replacing conventional ricks throughout Europe and are beginning to gain acceptance in North America as the quest to manage odors continues.

Europeans were the first to regulate emissions from their agricultural operations. Therefore, most European mushroom composting operations have employed some type of enclosed or environmentally controlled Phase I system. In North America, a few systems have been built to test the technology. Eventually they may become common at commercial operations. Unfortunately, little information is available to show how these systems reduce emissions.

Therefore, determining how effective aerated systems are in reducing odors is difficult.

Phase I is considered complete as soon as the raw ingredients become pliable and are capable of holding water, the odor of ammonia is sharp, and the dark-brown colour indicates that carmelization and browning reactions have occurred. At the beginning of Phase I, the substrate is bulky and yellow. At the end of Phase I substrate preparation, the substrate should be dense, chocolate brown in colour, and have a strong odor of ammonia. The substrate still has some structure so aeration can be maintained during Phase II composting. The potential fresh mushroom yield depends on the amount of dry weight filled. In order to achieve a substrate density in the growing structure necessary to support an economical mushroom yield, the substrate at fill has to be short or dense enough to attain a high substrate dry weight.

Growing Systems (Phase II)

Once Phase I is complete, the substrate will be filled into a system for Phase II substrate preparation and to grow the mushrooms. Phase II takes place in one of three main types of mushroom-growing systems, depending on the type of production system used. The difference in the mushroom- growing systems is the container in which the crop is processed and grown. Each room has a different heating, ventilating, and airconditioning (HVAC) system designed for a specific stage in crop development. A single-zone system—or bed farm—consists of several large, stacked beds or shelves within a single room. The substrate is filled into these beds after Phase I, and the crop remains in the one room throughout the process. Bulk pasteurization or tunnels are systems where the substrate is filled into "tractortrailer"– type bins (called tunnels) with perforated floors and no covers on top of the compost. Phase II and, occasionally, the next phase of growing are carried out within these tunnels. The substrate may then be filled into a tray, shelf, or even plastic garbage bags for the remaining part of the process.

Phase II: Finishing the Compost

Phase II composting is the second step of compost substrate preparation. The first objective of Phase II is to pasteurize the composted substrate. The composted substrate is pasteurized to reduce or eliminate the bad microbes such as insects, other fungi, and bacteria. This is not a complete sterilization but a selective killing of pests that will compete for food or directly attack the mushroom. At the same time, this process minimizes the loss of good microbes.

The second goal of Phase II is to complete the composting process. Completing the composting process means eliminating all remaining simple soluble sugars and gaseous and soluble ammonia created during Phase I composting. Since ammonia is toxic to the mushroom mycelium, it must be converted to food the mushroom can use. The good microbes in Phase II convert toxic ammonia in solution and amine (other readily available nitrogen compounds) substances into protein—specific food for the mushroom. At the end of Phase II, volatile ammonia (concentration more than 0.05 percent) will inhibit mushroom spawn growth. Generally, ammonia concentrations above 0.10 percent can be easily detected by a person and are toxic to the spawn. Most of this conversion of ammonia and carbohydrates is accomplished by the growth of the microbes in the compost. These microbes are very efficient in using Phase I composting products, such as ammonia, as one of their main sources of food. The ammonia is incorporated as mostly protein into their bodies or cells. Eventually the mushroom uses these packets of nutrients as food.

Phase II objectives are possibly the most difficult procedures in growing mushrooms. Because of a composting or other cultural problem, growers sometimes have to adjust Phase II programs. The Phase II process takes anywhere from 7 to 18 days, depending on how the air and compost temperatures are managed to control microbial activity.

During Phase II in standard bed or tray systems, compost temperatures are brought down through all temperature ranges to ensure that all the different species have a chance to convert their specific source of carbohydrates. The composted substrate throughout Phase II should appear to have moderate "firefang"—a term referring to the white-flecking microbial growth pattern of the thermophilic microorganism.

Pasteurization (peak heat, boost) should be completed toward the start of Phase II. Effective pasteurization will eradicate harmful bacteria, nematodes, insects, and fungi. In general, air and composted substrate temperatures should be raised together to 140°F (60°C) for at least 2 hours. Growers make several compromises to this range, but it is a time-temperature relationship.

The good microbes grow best at temperatures from 115°F to 140°F; the more ammonia-utilizing microbes grow best in the temperature range of 120–128°F (47–49°C). The longer the microbes in the composted substrate remain in this optimum range with all the critical growth requirements available, the faster the ammonia will be converted. Understanding how these microbes grow and work in composted substrate should make the management of Phase II a little easier. The process of going through this temperature range will produce the most protein or the maximum amount of food for the mushroom. A good rule of thumb is not to drop the composted substrate temperature more than 5°F per 24 hours, which maintains the compost substrate in the desired range for about 4 or more days.

Near the completion of Phase II, growers check for ammonia in the compost. The nose is usually the best tool. However, ammonia-testing kits and strips are available to supplement the nose test.

Spawn Maintenance

A desirable mycelial culture is pure—free of contaminants and of sectoring of other abnormalities. Contaminants include other fungi, bacteria, or insects growing on or infesting the culture media along with the desired mycelial culture. When a culture is first obtained, it should be transferred several times to fresh media to check for any form of contamination.

Sectoring is any type of mycelial growth that differs in appearance, growth rate, colour, or in any other way from the typical appearance of a given strain. Sectoring is often observed as a more rapidly growing area near the leading edge of growth, exhibiting a different growth habit from the rest of the culture. Other abnormalities that might appear in a culture are fluffy, aerial mycelia, thick or rubbery textures, and colour changes such as browning or darkening of the mycelium. Sectors of other change in vegetative growth could affect the productivity of the culture. Therefore, recognizing and avoiding propagation of abnormal mycelia to agar and further spawn production is very important.

There is no in vitro test to determine a stock culture's validity. A series of cropping trials must be conducted on the mycelial stock culture to determine a culture line's value. Mushroom yield, size, colour, cap shape, and any other desired quality or growth factors are selected and then compared for each culture line. Many commercially prepared spawn strains are available to commercial and noncommercial growers. All commercially grown strains are pure culture of edible, fresh mushrooms; some may vary in texture and growing requirements.

Mushroom spawn is produced in several different strains or isolates. Hybrid White is a smooth-cap, highyield, excellent processing strain. Hybrid Off-White has a cap that is slightly scaly on first break and is a preferred fresh-market strain, and Brown (Portabella, Crimini) produces a chocolate-brown, mature mushroom that is fleshy and has a strong, mature flavour.

Spawn Production

The process of making spawn remains much the same as Penn State professor emeritus Dr. Sinden first developed in the 1930s. Grain is mixed with a little calcium carbonate, then cooked, sterilized, and cooled. Small pieces of pure-culture mycelium are placed in small batches on the grain. Once the small batch is fully colonized, it is used to inoculate several larger batches of grain. This multiplying of the inoculated grain continues until the commercial-size containers— usually plastic bags with breathable filter patches—are inoculated. During the colonization of each batch, the containers are shaken every few days to distribute actively growing mycelia around the bag or bottle. During the process, temperatures are maintained at 74–76°F (23–24°C). Uniformity of the air circulating around the bags is important to ensure that all containers are kept within the desired temperature range. Mycelium is sensitive and its fruiting mechanism can be easily damaged at high temperatures.

Spawning

On bed farms, spawn and supplement are broadcast over the surface of the substrate. Uniformity of this distribution is critical to achieve even spawn growth and temperatures. On tray or bulk farms, spawn is usually metered into the substrate during the mixing operation. Spawning is the cleanest operation performed on a mushroom farm. All equipment, baskets, tools, and so forth should be thoroughly cleaned and disinfected before spawning.

The amount of spawn used depends on the length of the spawn-growing period and compost fill weights. The use of more spawn will result in a quicker colonization and more efficient use of substrate nutrients.

Improved colonization of substrate will help ensure that the mushroom mycelia will grow quicker than other fungal competitors.

During the spawn-growing period, heat is generated and supplemental cooling is required. Substrate temperatures should be maintained at 75–77°F and relative humidity should be high to minimize drying of the substrate surface. Under proper conditions, the spawn will grow as a delicate network of mycelia throughout the substrate. The mycelium grows in all directions from a spawn grain. Eventually mycelia from different spawn grains fuse together, making a spawned bed appear as a white root-like network throughout the compost. As the spawn grows, it generates heat. If the compost temperature increases to above 45 or 85°F, depending on the cultivar, the heat may kill or damage the mycelia, reducing crop yield and/or mushroom quality. The time needed for spawn to colonize the compost depends on the spawning rate and its distribution, the compost moisture and temperature, and the nature or quality of the compost. A complete spawn run usually requires 14 to 21 days. The spawn-growing period is considered complete when spawn has completely colonized the substrate and the metabolic heat surge is subsiding.

Substrate Supplementation

The compost has to provide the mushroom mycelium with a smorgasbord of food. Not only is ligninhumus complex and cellulose important, but protein, fat, and oils are also important. A good analogy is protein serves as the mushroom's "steak," carbohydrates its "potatoes," and lipids (fats and oils) its "butter." Like people, mushrooms should eat a balance of all these food types. The main source of "steak and butter" for the mushroom is from Phase II microbes. The dead cells of thermophilic fungi, bacteria, and actinomycetes "firefang" are the packages that deliver protein and fat to the mushroom. The addition of delayed-release supplements further enhances the protein and lipid content of the compost for the mushroom. Many of these supplements consist of a high-protein oil material, such as soybean meal, cornmeal, or feather meal, that has been treated to delay the availability of the nutrient for the mushroom. If an untreated supplement is added to the compost at this time, it often becomes a "candy bar" to other microbes,weeds, or competitor molds. These molds grow more rapidly than the mushroom mycelium and can quickly colonize the compost, competing with the mushroom for nutrients. The oils or lipids in these supplements are used by the mushroom to stimulate the fruiting mechanism and increase yield by having more mushrooms initiate and develop. Yields can be increased from 0.25 to 1.5 lbs/sq ft of growing space. In addition, mushroom size may also be improved in compost with higher spawning-moisture content. However, in substrate that is not selectively prepared, these nutrients become more available to com petitor molds. Often, if a farm is having composting problems, not supplementing until the problems are corrected is more economical.

Casing

The only method of forcing mushroom mycelia to change from the vegetative phase to a reproductive state is to apply a cover of a suitable material—called the casing layer—on the surface of the spawned compost.

The function of a casing layer is to trigger the mushrooms to switch from a vegetative growth to a reproductive or fruiting growth. The mechanism that initiates the spawn to change from vegetative to reproductive growth is unknown, though several theories have been presented. The casing also functions to supply and conserve moisture for the mushrooms and their rhizomorphs (thicker mushroom mycelia) and acts to transport dissolved nutrients to the mushrooms. Casing supports the mushrooms and compensates for water lost through evaporation and transpiration. Rhizomorphs look like thick strings. They are formed when the very fine mycelia fuse together and grow through the casing.

Rhizomorphs are thought to carry water and nutrients from the compost to the developing mushrooms. Mushroom initials—primordia or pins—form on the rhizomorphs. Without rhizomorphs, there will be no mushrooms.

The mushroom industry uses various materials to provide a suitable environment for fruit body formations. Presently, most mushroom growers use sphagnum peat moss or aged sphagnum peat moss buffered with limestone. Sphagnum peat is relatively inexpensive and readily available to North American growers. Pasteurized clay loam field soil; reclaimed, weathered, spent compost; and coir fibres are other materials used by growers.

Most sphagnum peat has a pH of 3.5 to 4.5. A neutralizing agent—usually calcium limestone—is added to bring the pH level up to 7.5. Processed, spent sugar beet lime or hydrated lime can be used. Due to its higher neutralizing capability and its greater solubility, only small amounts are required.

Soil, spent mushroom substrate, and coir fibres should be pasteurized to eliminate any insects and pathogens they may be carrying. However, peat moss–based casing does not need pasteurization because it is inherently free of mushroom disease spores and pests. Distributing the casing so the depth and moisture are uniform over the surface of the compost is important.

Such uniformity allows spawns to move into and through the casing at the same rate and, ultimately, for mushrooms to develop at the same time. Casing should be able to hold moisture because moisture is essential for the development of a firm mushroom.

CAC or CI

Fully colonized spawn run substrate is used to introduce mycelia into the casing layer. This is often used to improve crop uniformity, crop cycling, mushroom quality, and yields. Spawn run compost at casing (CAC) is used to inoculate the casing during the mixing or application of the casing. CAC is now produced much like spawn—in aseptic conditions—by those who produce and supply spawn to growers. This process is called casing inoculum (CI). By adding the mycelia uniformly throughout the casing, the spawn growth into the casing is quicker and more even. The time from casing to harvest is reduced by 5–7 days so that the rooms can be cycled faster or more breaks can be harvested in the same time. Mycelial growth is uniform on the surface, which encourages the mushrooms to form on the surface as well. Therefore, they are cleaner. Yields are improved since the mushroom growth is uniform and crop management is easier. In addition, more mushrooms are produced from areas that may have less nutrition.

Managing the crop after casing requires that the compost temperatures until flushing be held at spawngrowing temperatures. After flushing, compost temperatures are lowered and air temperature becomes the primary control point. Throughout the period following casing, water must be applied intermittently to raise the moisture level to field capacity before the mushroom pins form.

Watering or Irrigation

The moisture content of the casing often determines the uniformity of the casing depth. Casing, both by equipment and by hand, becomes more difficult as the casing material increases in moisture. Peat moss casing will lump up or adhere to the different parts of the equipment, making the flow of the material uneven. Knowing when, how, and how much water to apply to casing is an art form that readily separates experienced growers from beginners. Watering the crop is the most delicate operation in mushroom growing. Although each grower may have his or her own preference, no specific casing-management practice and casing material are universally accepted. Despite so much diversity, many growers are still able to harvest good crops with good freshmarket quality.

Although much has been written about when and how much water to apply at certain stages in the crop's development, most growers rely on their ability to "read" the crop and determine how the mushrooms look from day to day. Water constantly moves throughout the cropping period: water is lost through evaporation and transpiration, and the mushroom takes up water into its cells; water is replaced when watering the casing layer. The increase in the weight of the mushroom from pinning to maturing is related to the rapid uptake of water from the casing and compost. The mushroom doubles in size 2 days before harvest, putting more strain on the pipe system in the compost and casing. As the mushroom matures during a flush, its weight gain is attributed to the accumulation of nutrients and water from the substrate.

Pinning

Mushroom initials develop after rhizomorphs have formed in the casing. The initials are extremely small but can be seen as clumps on a rhizomorph. As these structures grow and expand, they are called primordial or pins. Mushroom pins continue to grow larger through a prebutton stage and ultimately enlarge to mature mushrooms. Mushroom harvesting begins 15–21 days after casing, which is normally 10–12 days after flushing and 7–8 weeks after composting started. The cultural practices used during pin development and cropping include the management of air and compost temperatures and CO2 content of room air, and is often dependent on the strain and number of pins the grower wishes to form and develop.

Air-handling systems regulate the amount of fresh air entering the room and temperatures within the room. Ventilation requirements depend on the amount of mushrooms to be grown on the beds, heat, and CO_2 production, which increases with temperature. Uniform air movement and circulation is important to prevent stale air with high CO2 levels from building up around the mushrooms, which lowers fresh quality. Air temperature is maintained in a range of 60–66°F (15–17°C); CO2 levels range from 1,000 to 2,500 ppm (1–2.5 percent) during the pinning and cropping stages. The most critical stage of the mushroom's development for fresh quality and yield improvement is during the Rapidly Expanding Stage (RES), when the mushroom doubles in size every 24 hours.

This expansion stage depends on temperature, moisture of the compost, and casing. The environment inside a production room determines the rate of transpiration, which aids in the flow of nutrients and moisture into the mushrooms.

Mushroom size is dependent on the number of pins that develop for a break or flush and by how the crop is prepared and managed. Portabella mushroom growers have learned to manage the pin set to achieve enough pins for good yield, yet, more important, to attain the right amount of pins to produce the large mature mushrooms for the Portabella market.

Harvesting

Mushrooms are harvested over a 2–4- day period in a 7–10-day cycle called flushes or breaks. When mature mushrooms are picked, an inhibitor to mushroom development is removed and the next flush moves toward maturity. Timing of the breaks or flushes is managed by control of the watering, CO2, and temperatures. The first two flushes account for the majority of the total yield, with the subsequent flushes tapering off to relatively low levels of production.

Mushrooms are harvested by hand and are picked at a time before the cap becomes soft, indicating the mushroom is past prime fresh-quality potential. Harvesting rates depend mainly on the amount of crop on the beds and size of the mushrooms.

Rates vary from 30 to 80 lbs/hour. Some consumers seem to prefer closed, tight mushrooms, while others prefer stronger-flavoured, more mature, open-cap mushrooms. Mushroom maturity is evaluated by how open the veil is, not by its size. Mature mushrooms are both large and small, although both farmers and consumers favour medium to large mushrooms.

Growers harvest just three to four breaks per crop—a shorter harvesting time allows more crops to be produced in a year and helps to prevent disease and insect problems.

Diseased, malformed, and fly-damaged mushrooms are considered second-grade and are discarded.

Diseased mushrooms should not be touched. Diseased tissue should be treated with registered chemicals, biopesticides, or common disinfectant materials such as salt or alcohol.

6

Biosynthesis of Cellulose and Bioconversion of Cellulosic Wastes

Cellulose is an organic compound with the formula $(C_6H_{10}O_5)_n$, a polysaccharide consisting of a linear chain of several hundred to over ten thousand $\beta(1\rightarrow4)$ linked D-glucose units.

Cellulose is the structural component of the primary cell wall of green plants, many forms of algae and the oomycetes. Some species of bacteria secrete it to form biofilms. Cellulose is the most common organic compound on Earth. About 33% of all plant matter is cellulose (the cellulose content of cotton is 90% and that of wood is 40–50%).

For industrial use, cellulose is mainly obtained from wood pulp and cotton. It is mainly used to produce paperboard and paper; to a smaller extent it is converted into a wide variety of derivative products such as cellophane and rayon. Converting cellulose from energy crops into biofuels such as cellulosic ethanol is under investigation as an alternative fuel source.

Some animals, particularly ruminants and termites, can digest cellulose with the help of symbiotic micro-organisms that live in their guts. Humans can digest cellulose to some extent, however it is often referred to as 'dietary fiber' or 'roughage' (e.g. outer shell of maize) and acts as a hydrophilic bulking agent for feces.

History

Cellulose was discovered in 1838 by the French chemist Anselme Payen, who isolated it from plant matter and determined its chemical formula. Cellulose was used to produce the first successful thermoplastic polymer, celluloid, by Hyatt Manufacturing Company in 1870. Hermann Staudinger determined the polymer structure of cellulose in 1920. The compound was first chemically synthesized (without the use of any biologically derived enzymes) in 1992, by Kobayashi and Shoda.

Commercial Products

Cellulose is the major constituent of paper, paperboard, and card stock and of textiles made from cotton, linen, and other plant fibres.

Cellulose can be converted into cellophane, a thin transparent film, and into rayon, an important fibre that has been used for textiles since the beginning of the 20th century. Both cellophane and rayon are known as "regenerated cellulose fibres"; they are identical to cellulose in chemical structure and are usually made from dissolving pulp via viscose. A more recent and environmentally friendly method to produce rayon is the Lyocell process. Cellulose is the raw material in the manufacture of nitrocellulose (cellulose nitrate) which was historically used in smokeless gunpowder and as the base material for celluloid used for photographic and movie films until the mid 1930s.

Cellulose is used to make water-soluble adhesives and binders such as methyl cellulose and carboxymethyl cellulose which are used in wallpaper paste. Microcrystalline cellulose (E460i) and powdered cellulose (E460ii) are used as inactive fillers in tablets and as thickeners and stabilizers in processed foods. Cellulose powder is for example used in Kraft's Parmesan cheese to prevent caking inside the tube.

Cellulose is used in the laboratory as the stationary phase for thin layer chromatography. Cellulose fibres are also used in liquid filtration, sometimes in combination with diatomaceous earth or other filtration media, to create a filter bed of inert material. Cellulose is further used to make hydrophilic and highly absorbent sponges.

Cellulose insulation made from recycled paper is becoming popular as an environmentally preferable material for building insulation. It can be treated with boric acid as a fire retardant.

Cellulose Source and Energy Crops

The major combustible component of non-food energy crops is cellulose, with lignin second. Non-food energy crops are more efficient than edible energy crops (which have a large starch component), but still compete with food crops for agricultural land and water resources. Typical non-food energy crops include industrial hemp, switchgrass, *Miscanthus*, *Salix* (willow), and *Populus* (poplar) species.

Structure and Properties

Cellulose has no taste, is odourless, is hydrophilic with the contact angle of 20–30, is insoluble in water and most organic solvents, is chiral and is biodegradable. It can be broken down chemically into its glucose units by treating it with concentrated acids at high temperature.

Cellulose is derived from D-glucose units, which condense through β(1→4)-glycosidic bonds. This linkage motif contrasts with that for α(1→4)-glycosidic bonds present in starch, glycogen, and other carbohydrates. Cellulose is a straight chain polymer: unlike starch, no coiling or branching occurs, and the molecule adopts an extended and rather stiff rod-like conformation, aided by the equatorial conformation of the glucose residues. The multiple hydroxyl groups on the glucose from one chain form hydrogen bonds with oxygen molecules on the same or on a neighbour chain, holding the chains firmly together side-by-side and forming *microfibrils* with high tensile strength. This strength is important in cell walls, where the microfibrils are meshed into a carbohydrate *matrix*, conferring rigidity to plant cells.

Compared to starch, cellulose is also much more crystalline. Whereas starch undergoes a crystalline to amorphous transition when heated beyond 60-70 °C in water (as in cooking), cellulose requires a temperature of 320 °C and pressure of 25 MPa to become amorphous in water.

Several different crystalline structures of cellulose are known, corresponding to the location of hydrogen bonds between and within strands. Natural cellulose is cellulose I, with structures Iα and Iβ. Cellulose produced by bacteria and algae is enriched in Iα while cellulose of higher plants consists mainly of Iβ. Cellulose in regenerated cellulose fibres is cellulose II. The conversion of cellulose I to cellulose II is not reversible, suggesting that cellulose I is metastable and cellulose II is stable. With various chemical treatments it is possible to produce the structures cellulose III and cellulose IV. Many properties of cellulose depend on its chain length or degree of polymerization, the number of glucose units that make up one polymer molecule. Cellulose from wood pulp has typical chain lengths between 300 and 1700 units; cotton and other plant fibres as well as bacterial celluloses have chain lengths ranging from 800 to 10,000 units. Molecules with very small chain length resulting from the breakdown of cellulose are known as cellodextrins; in contrast to long-chain cellulose, cellodextrins are typically soluble in water and organic solvents.

Plant-derived cellulose is usually found in a mixture with hemicellulose, lignin, pectin and other substances, while microbial cellulose is quite pure, has a much higher water content, and consists of long chains.

Cellulose is soluble in cupriethylenediamine (CED), cadmiumethylenediamine (Cadoxen), *N*-methylmorpholine *N*-oxide and lithium chloride/dimethylformamide. This is used in the production of regenerated celluloses (as viscose and cellophane) from dissolving pulp.

Assaying a Cellulose-containing Material

Given a cellulose-containing material, the carbohydrate portion that does not dissolve in a 17.5% solution of sodium hydroxide at 20 °C is α *cellulose*, which is true cellulose. Acidification of the extract precipitates β *cellulose*. The portion that dissolves in base but does not precipitate with acid is γ *cellulose*.

Cellulose can be assayed using a method described by Updegraff in 1969, where the fibre is dissolved in acetic and nitric acid to remove lignin, hemicellulose, and xylosans. The resulting cellulose is allowed to react with anthrone in sulfuric acid. The resulting coloured compound is assayed spectrophotometrically at a wavelength of approximately 635 nm.

In addition, cellulose is represented by the difference between acid detergent fibre (ADF) and acid detergent lignin (ADL).

Biosynthesis

In vascular plants cellulose is synthesized at the plasma membrane by rosette terminal complexes (RTCs). The RTCs are hexameric protein structures, approximately 25 nm in diameter, that contain the cellulose synthase enzymes that synthesise the individual cellulose chains. Each RTC floats in the cell's plasma membrane and "spins" a microfibril into the cell wall.

RTCs contain at least three different cellulose synthases, encoded by *CesA* genes, in an unknown stoichiometry. Separate sets of *CesA* genes are involved in primary and secondary cell wall biosynthesis.

Cellulose synthesis requires chain initiation and elongation, and the two processes are separate. *CesA* glucosyltransferase initiates cellulose polymerization using a steroid primer, sitosterol-beta-glucoside, and UDP-glucose. Cellulose synthase utilizes UDP-D-glucose precursors to elongate the growing cellulose chain. A cellulase may function to cleave the primer from the mature chain.

Breakdown (Cellulolysis)

Cellulolysis is the process of breaking down cellulose into smaller polysaccharides called cellodextrins or completely into glucose units; this is a hydrolysis reaction. Because cellulose molecules bind strongly to each other, cellulolysis is relatively difficult compared to the breakdown of other polysaccharides. Processes do exist however for the breakdown of cellulose such as the Lyocell process which uses a combination of heated water and acetone to break down the cellulose strands.

Most mammals have only very limited ability to digest dietary fibres such as cellulose. Some ruminants like cows and sheep contain certain symbiotic anaerobic bacteria (like *Cellulomonas*) in the flora of the rumen, and these bacteria produce enzymes called cellulases that help the microorganism to break down cellulose; the breakdown products are then used by the bacteria for proliferation. The bacterial mass is later digested by the ruminant in its digestive system (stomach and small intestine). Similarly, lower termites contain in their hindguts certain flagellate protozoa which produce such enzymes; higher termites contain bacteria for the job. Some termites may also produce cellulase of their own. Fungi, which in nature are responsible for recycling of nutrients, are also able to break down cellulose.

The enzymes utilized to cleave the glycosidic linkage in cellulose are glycoside hydrolases including endo-acting cellulases and exo-acting glucosidases. Such enzymes are usually secreted as part of multienzyme complexes that may include dockerins and cellulose binding modules.

Hemicellulose

Hemicellulose is a polysaccharide related to cellulose that comprises ca. 20% of the biomass of most plants. In contrast to cellulose, hemicellulose is derived from several sugars in addition to glucose, especially xylose but also including mannose, galactose, rhamnose, and arabinose. Hemicellulose consists of shorter chains- around 200 sugar units. Furthermore, hemicellulose is branched, whereas cellulose is unbranched.

Derivatives

The hydroxyl groups (-OH) of cellulose can be partially or fully reacted with various reagents to afford derivatives with useful properties like mainly cellulose esters and cellulose ethers (-OR). In principle, though not always in current industrial practice, cellulosic polymers are renewable resources.

Microbial Cellulose

Microbial cellulose is a form of cellulose that is produced by bacteria. It is widely used in the traditional Filipino dessert Nata de coco. Microbial cellulose was first confirmed as cellulose in 1886.

Production

Bacteria from the genera *Aerobacter*, *Acetobacter*, *Achromobacter*, *Agrobacterium*, *Alacaligenes*, *Azotobacter*, *Pseudomonas*, *Rhizobium* and *Sarcina* synthesize cellulose. However, only the *Acetobacter* species produce enough cellulose to justify commercial interest. The most extensively studied member of the *Acetobacter* species is *A. xylinus*, formerly known as *A. xylinum*.

A.xylinus extrudes glucan chains from pores into the growth medium. These aggregate into microfibrils, which bundle to form microbial cellulose ribbons. Various kinds of sugars are used as substrate. Production occurs mostly at the interface of liquid and air.

Differences with Plant Cellulose

Some advantages of microbial cellulose over plant cellulose include:

- Finer and more intricate structure
- No hemicellulose or lignin need to be removed
- Longer fibre length: much stronger
- Can be grown to virtually any shape
- Can be produced on a variety of substrates
- The formula of the media used and the strain of Acetobacter *xylinum* will determine the quality of the pellicle
- More absorbent per unit volume.

Disadvantages for Commercial Use

Some issues that prevented larger scale commercialization so far include:

- High price (about 100 x more than plant cellulose)
    - o Because of high priced substrates: sugars
    - o Low volumetric yields
- Lack of large scale production capacity
- Timely expansion and maintenance of the cell culture for production.

Function

One continuing mystery surrounding microbial cellulose is its exact biological function. *A. xylinus* is a successful and prevalent bacterium in nature, frequently finding a home in rotting fruits and sweetened liquids. The most familiar form of microbial cellulose is that of a pellicle on the top of a static cultured growth media. It has thus been hypothesized that cellulose acts as a floatation device bringing the bacteria to the oxygen rich air-media interface. This hypothesis has largely been discredited by experiments conducted on submerged oxygen-permeable silicone tubes that show cellulose grows well submerged if enough oxygen is present. Others suspect that cellulose is used to immobilize the bacteria in an attempt to keep it near the food source; or as a form of protection against ultraviolet light.

Uses

Medical

Microbial cellulose is biocompatible and non-toxic making it a good candidate material for medical applications. So far it has found a commercial role in some wound dressings. There is on-going research to evaluate a possible role for bacterial cellulose in the following applications:

- Scaffolds for tissue engineering
- Synthetic dura mater
- Bladder neck suspension
- Soft tissue replacement
- Artificial blood vessels.

Non Medical

- Matrix for electronic paper
- High strength paper
- Diet foods
- Desserts: nata de coco
- Substrates for OLEDs
- Sony has use microbial cellulose as an acoustic membrane in high-end earphones
- It has a proposed use as a gloss surface finish in magazines.

Management of Soil by Sustainable Utilization of Resources

Soil management concerns all operations, practices and treatments used to protect soil and enhance its performance.

What are some features of good soil? Any farmer will tell you that a good soil:

- feels soft and crumbles easily
- drains well and warms up quickly in the spring
- does not crust after planting
- soaks up heavy rains with little runoff
- stores moisture for drought periods
- has few clods and no hardpan
- resists erosion and nutrient loss
- supports high populations of soil organisms
- has a rich, earthy smell
- does not require increasing inputs for high yields
- produces healthy, high-quality crops.

All these criteria indicate a soil that functions effectively today and will continue to produce crops long into the future. These characteristics can be created through management practices that optimize the processes found in native soils.

How does soil in its native condition function? How do forests and native grasslands produce plants and animals in the complete absence of fertilizer and tillage? Understanding the principles by which native soils function can help farmers develop and maintain productive and profitable soil both now and for future generations. The soil, the environment, and farm condition benefit when the soil's natural productivity is managed in a sustainable way. Reliance on purchased inputs declines year by year, while land value and income potential increase. Some of the things we spend money on can be done by the natural process itself for little or nothing. Good soil management produces crops and animals that are healthier, less susceptible to disease, and more productive. To understand this better, let's start with the basics.

Sustainable: capable of being maintained at length without interruption, weakening, or losing in power or quality.

The Living Soil: Texture and Structure

Soils are made up of four basic components: minerals, air, water, and organic matter. In most soils, minerals represent around 45% of the total volume, water and air about 25% each, and organic matter from 2% to 5%. The mineral portion consists of three distinct particle sizes classified as sand, silt, or clay. Sand is the largest particle that can be considered soil.

Sand is largely the mineral quartz, though other minerals are also present. Quartz contains no plant nutrients, and sand cannot hold nutrients—they leach out easily with rainfall. Silt particles are much smaller than sand, but like sand, silt is mostly quartz. The smallest of all the soil particles is clay. Clays are quite different from sand or silt, and most types of clay contain appreciable amounts of plant nutrients. Clay has a large surface area resulting from the plate-like shape of the individual particles. Sandy soils are less productive than silts, while soils containing clay are the most productive and use fertilizers most effectively.

Soil texture refers to the relative proportions of sand, silt, and clay. A loam soil contains these three types of soil particles in roughly equal proportions. A sandy loam is a mixture containing a larger amount of sand and a smaller amount of clay, while a clay loam contains a larger amount of clay and a smaller amount of sand.

The Living Soil: The Importance of Soil Organisms

An acre of living topsoil contains approximately 900 pounds of earthworms, 2,400 pounds of fungi, 1,500 pounds of bacteria, 133 pounds of protozoa, 890 pounds of arthropods and algae, and even small mammals in some cases. Therefore, the soil can be viewed as a living community rather than an inert body. Soil organic matter also contains dead organisms, plant matter, and other organic materials in various phases of decomposition. Humus, the dark-coloured organic material in the final stages of decomposition, is relatively stable. Both organic matter and humus serve as reservoirs of plant nutrients; they also help to build soil structure and provide other benefits. The type of healthy living soil required to support humans now and far into the future will be balanced in nutrients and high in humus, with a broad diversity of soil organisms. It will produce healthy plants with minimal weed, disease, and insect pressure. To accomplish this, we need to work *with* the natural processes and optimize their functions to sustain our farms.

Considering the natural landscape, you might wonder how native prairies and forests function in the absence of tillage and fertilizers. These soils are tilled by soil organisms, not by machinery. They are fertilized too, but the fertility is used again and again and never leaves the site. Native soils are covered with a layer of plant litter and/or growing plants throughout the year. Beneath the surface litter, a rich complexity of soil organisms decompose plant residue and dead roots, then release their stored nutrients slowly over time. In fact, topsoil is the most biologically diverse part of the earth. Soil-dwelling organisms release bound-up minerals, converting them into plant-available forms that are then taken up by the plants growing on the site. The organisms recycle nutrients again and again with the death and decay of each new generation of plants. There are many different types of creatures that live on or in the topsoil. Each has a role to play. These organisms will work for the farmer's benefit if we simply manage for their survival. Consequently, we may refer to them as soil livestock. While a great variety of organisms contribute to soil fertility, earthworms, arthropods, and the various microorganisms merit particular attention.

Earthworms

Earthworm burrows enhance water infiltration and soil aeration. Fields that are "tilled" by earthworm tunneling can absorb water at a rate 4 to 10 times that of fields lacking worm tunnels. This reduces water runoff, recharges groundwater, and helps store more soil water for dry spells. Vertical earthworm burrows pipe air deeper into the soil, stimulating microbial nutrient cycling at those deeper levels. When earthworms are present in high numbers, the tillage provided by their burrows can replace some expensive tillage work done by machinery.

Worms eat dead plant material left on top of the soil and redistribute the organic matter and nutrients throughout the topsoil layer. Nutrient-rich organic compounds line their tunnels, which may remain in place for years if not disturbed. During droughts these tunnels allow for deep plant root penetration into subsoil regions of higher moisture content. In addition to organic matter, worms also consume soil and soil microbes. The soil clusters they expel from their digestive tracts are known as worm casts or castings. These range from the size of a mustard seed to that of a sorghum seed, depending on the size of the worm.

The soluble nutrient content of worm casts is considerably higher than that of the original soil. A good population of earthworms can process 20,000 pounds of topsoil per year—with turnover rates as high as 200 tons per acre having been reported in some exceptional cases. Earthworms also secrete a plant growth stimulant. Reported increases in plant growth following earthworm activity may be partially attributed to this substance, not just to improved soil quality.

Table : Selected nutrient analyses of worm casts compared to those of the surrounding soil.

Nutrient	*Worm casts (Lbs/ac)*	*Soil (Lbs/ac)*
Carbon	171,000	78,500
Nitrogen	10,720	7,000
Phosphorus	280	40
Potassium	900	140

From Graff. Soil had 4% organic matter.

Earthworms thrive where there is no tillage. Generally, the less tillage the better, and the shallower the tillage the better. Worm numbers can be reduced by as much as 90% by deep and frequent tillage. Tillage reduces earthworm populations by drying the soil, burying the plant residue they feed on, and making the soil more likely to freeze. Tillage also destroys vertical worm burrows and can kill and cut up the worms themselves. Worms are dormant in the hot part of the summer and in the cold of winter. Young worms emerge in spring and fall-they are most active just when farmers are likely to be tilling the soil.

Table : Effect of crop Management on earthworm populations.

Crop	*Management*	*Worms/foot^2*
Corn	Plow	1
Corn	No-till	2
Soybean	Plow	6
Soybean	No-till	14
Bluegrass/clover	—	39
Dairy pasture	—	33

From Kladivko

As a rule, earthworm numbers can be increased by reducing or eliminating tillage (especially fall tillage), not using a moldboard plow, reducing residue particle size (using a straw chopper on the combine), adding animal manure, and growing green manure crops. It is beneficial to leave as much surface residue as possible year-round. Cropping systems that typically have the most earthworms are (in descending order) perennial cool-season grass grazed rotationally, warm-season perennial grass grazed rotationally, and annual croplands using no-till. Ridge-till and strip tillage will generally have more earthworms than clean tillage involving plowing and disking. Cool season grass rotationally grazed is highest because it provides an undisturbed (no-tillage) environment plus abundant organic matter from the grass roots and fallen grass litter. Generally speaking, worms want their food on top, and they want to be left alone.

Earthworms prefer a near-neutral soil pH, moist soil conditions, and plenty of plant residue on the soil surface. They are sensitive to certain pesticides and some incorporated fertilizers. Carbamate insecticides, including Furadan, Sevin, and Temik, are harmful to earthworms, notes worm biologist Clive Edwards of Ohio State University. Some insecticides in the organophosphate family are mildly toxic to earthworms, while synthetic pyrethroids are harmless to them. Most herbicides have little effect on worms except for the triazines, such as Atrazine, which are moderately toxic. Also, anhydrous ammonia kills earthworms in the injection zone because it dries the soil and temporarily increases the pH there. High rates of ammonium-based fertilizers are also harmful.

For more information on managing earthworms, order *The Farmer's Earthworm Handbook: Managing Your Underground Moneymakers*, by David Ernst. Ernst's book contains details on what earthworms need to live, how to increase worm numbers, the effects of tillage, manure, and livestock management on earthworms, how 193 chemicals affect earthworms, and more.

As a rule, earthworm numbers can be increased by reducing or eliminating tillage.

Arthropods

In addition to earthworms, there are many other species of soil organisms that can be seen by the naked eye. Among them are sowbugs, millipedes, centipedes, slugs, snails, and springtails. These are the primary decomposers. Their role is to eat and shred the large particles of plant and animal residues.

Some bury residue, bringing it into contact with other soil organisms that further decompose it. Some members of this group prey on smaller soil organisms. The springtails are small insects that eat mostly fungi.

Their waste is rich in plant nutrients released after other fungi and bacteria decompose it. Also of interest are dung beetles, which play a valuable role in recycling manure and reducing livestock intestinal parasites and flies.

Bacteria

Bacteria are the most numerous type of soil organism: every gram of soil contains at least a million of these tiny one-celled organisms. There are many different species of bacteria, each with its own role in the soil environment. One of the major benefits bacteria provide for plants is in making nutrients available to them. Some species release nitrogen, sulfur, phosphorus, and trace elements from organic matter. Others break down soil minerals, releasing potassium, phosphorus, magnesium, calcium, and iron. Still other species make and release plant growth hormones, which stimulate root growth. Several species of bacteria transform nitrogen from a gas in the air to forms available for plant use, and from these forms back to a gas again. A few species of bacteria fix nitrogen in the roots of legumes, while others fix nitrogen independently of plant association. Bacteria are responsible for converting nitrogen from ammonium to nitrate and back again, depending on certain soil conditions. Other benefits to plants provided by various species of bacteria include increasing the solubility of nutrients, improving soil structure, fighting root diseases, and detoxifying soil.

Fungi

Fungi come in many different species, sizes, and shapes in soil. Some species appear as thread-like colonies, while others are one-celled yeasts. Slime molds and mushrooms are also fungi. Many fungi aid plants by breaking down organic matter or by releasing nutrients from soil minerals. Fungi are generally quick to colonize larger pieces of organic matter and begin the decomposition process. Some fungi produce plant hormones, while others produce antibiotics including penicillin. There are even species of fungi that trap harmful plant-parasitic nematodes.

The mycorrhizae (my-cor-ry´-zee) are fungi that live either on or in plant roots and act to extend the reach of root hairs into the soil. Mycorrhizae increase the uptake of water and nutrients, especially phosphorus. They are particularly important in degraded or less fertile soils. Roots colonized by mycorrhizae are less likely to be penetrated by root-feeding nematodes, since the pest cannot pierce the thick fungal network. Mycorrhizae also produce hormones and antibiotics that enhance root growth and provide disease suppression. The fungi benefit by taking nutrients and carbohydrates from the plant roots they live in.

Actinomycetes

Actinomycetes (ac-tin-o-my´-cetes) are thread-like bacteria that look like fungi. While not as numerous as bacteria, they too perform vital roles in the soil. Like the bacteria, they help decompose organic matter into humus, releasing nutrients. They also produce antibiotics to fight diseases of roots. Many of these same antibiotics are used to treat human diseases. Actinomycetes are responsible for the sweet, earthy smell noticed whenever a biologically active soil is tilled.

Algae

Many different species of algae live in the upper half-inch of the soil. Unlike most other soil organisms, algae produce their own food through photosynthesis. They appear as a greenish film on the soil surface following a saturating rain. Algae improve soil structure by producing slimy substances that glue soil together into water-stable aggregates. Some species of algae (the blue-greens) can fix their own nitrogen, some of which is later released to plant roots.

Protozoa

Protozoa are free-living microorganisms that crawl or swim in the water between soil particles. Many soil protozoa are predatory, eating other microbes. One of the most common is an amoeba that eats bacteria. By eating and digesting bacteria, protozoa speed up the cycling of nitrogen from the bacteria, making it more available to plants.

Nematodes

Nematodes are abundant in most soils, and only a few species are harmful to plants. The harmless species eat decaying plant litter, bacteria, fungi, algae, protozoa, and other nematodes. Like other soil predators, nematodes speed the rate of nutrient cycling.

Soil Organisms and Soil Quality

All these organisms—from the tiny bacteria up to the large earthworms and insects—interact with one another in a multitude of ways in the soil ecosystem. Organisms not directly involved in decomposing plant wastes may feed on each other or each other's waste products or the other substances they release. Among the substances released by the various microbes are vitamins, amino acids, sugars, antibiotics, gums, and waxes.

Research on life in the soil has determined that there are ideal ratios for certain key organisms in highly productive soils.

Roots can also release into the soil various substances that stimulate soil microbes. These substances serve as food for select organisms. Some scientists and practitioners theorize that plants use this means to stimulate the specific population of microorganisms capable of releasing or otherwise producing the kind of nutrition needed by the plants.

Research on life in the soil has determined that there are ideal ratios for certain key organisms in highly productive soils. The Soil Foodweb Lab, located in Oregon, tests soils and makes fertility recommendations that are based on this understanding. Their goal is to alter the makeup of the soil microbial community so it resembles that of a highly fertile and productive soil. There are several different ways to accomplish this goal, depending on the situation. For more on the Soil Foodweb Lab, see the Additional Resources section of this publication.

Because we cannot see most of the creatures living in the soil and may not take time to observe the ones we can see, it is easy to forget about them. There are many Web sites that provide in-depth information on soil organisms. Look for a list of these Web sites in the Additional Resources section. Many of these sites have colour photographs of soil organisms and describe their benefits to soil fertility and plant growth.

Table : Weights of soil organisms in the top 7 inches of fertile soil.

Organism	*Pounds of liveweight/acre*
Bacteria	1000
Actinomycetes	1000
Molds	2000
Algae	100
Protozoa	200
Nematodes	50
Insects	100
Worms	1000
Plant roots	2000

From Bollen.

Organic Matter, Humus, and the Soil Foodweb

Understanding the role that soil organisms play is critical to sustainable soil management. Based on that understanding, focus can be directed toward strategies that build both the numbers and the diversity of soil organisms. Like cattle and other farm animals, soil livestock require proper feed. That feed comes in the form of organic matter.

Like cattle and other farm animals, soil livestock require proper feed.

Organic matter and humus are terms that describe somewhat different but related things. Organic matter refers to the fraction of the soil that is composed of both living organisms and once-living residues in various stages of decomposition. Humus is only a small portion of the organic matter. It is the end product of organic matter decomposition and is relatively stable. Further decomposition of humus occurs very slowly in both agricultural and natural settings. In natural systems, a balance is reached between the amount of humus formation and the amount of humus decay. This balance also occurs in most agricultural soils, but often at a much lower level of soil humus. Humus contributes to well-structured soil that, in turn, produces high-quality plants. It is clear that management of organic matter and humus is essential to sustaining the whole soil ecosystem.

The benefits of a topsoil rich in organic matter and humus are many. They include rapid decomposition of crop residues, granulation of soil into water-stable aggregates, decreased crusting and clodding, improved internal drainage, better water infiltration, and increased water and nutrient holding capacity. Improvements in the soil's physical structure facilitate easier tillage, increased water storage capacity, reduced erosion, better formation and harvesting of root crops, and deeper, more prolific plant root systems.

Soil organic matter can be compared to a bank account for plant nutrients. Soil containing 4% organic matter in the top seven inches has 80,000 pounds of organic matter per acre. That 80,000 pounds of organic matter will contain about 5.25% nitrogen, amounting to 4,200 pounds of nitrogen per acre. Assuming a 5% release rate during the growing season, the organic matter could supply 210 pounds of nitrogen to a crop. However, if the organic matter is allowed to degrade and lose nitrogen, purchased fertilizer will be necessary to prop up crop yields.

All the soil organisms mentioned previously, except algae, depend on organic matter as their food source. Therefore, to maintain their populations, organic matter must be renewed from plants growing on the soil, or from animal manure, compost, or other materials imported from off site. When soil livestock are fed, fertility is built up in the soil, and the soil will feed the plants.

Ultimately, building organic matter and humus levels in the soil is a matter of managing the soil's living organisms—something akin to wildlife management or animal husbandry. This entails working to maintain favourable conditions of moisture, temperature, nutrients, pH, and aeration. It also involves providing a steady food source of raw organic material.

Soil Tilth and Organic Matter

A soil that drains well, does not crust, takes in water rapidly, and does not make clods is said to have good tilth. Tilth is the physical condition of the soil as it relates to tillage ease, seedbed quality, easy seedling emergence, and deep root penetration. Good tilth is dependent on aggregation—the process whereby individual soil particles are joined into clusters or "aggregates."

Ultimately, building organic matter and humus in the soil is a matter of managing the soil's living organisms.

Aggregates form in soils when individual soil particles are oriented and brought together through the physical forces of wetting and drying or freezing and thawing. Weak electrical forces from calcium and magnesium hold soil particles together when the soil dries. When these aggregates become wet again, however, their stability is challenged, and they may break apart. Aggregates can also be held together by plant roots, earthworm activity, and by glue-like products produced by soil microorganisms. Earthworm-created aggregates are stable once they come out of the worm. An aggregate formed by physical forces can be bound together by fine root hairs or threads produced by fungi.

Aggregates can also become stabilized (remain intact when wet) through the by-products of organic matter decomposition by fungi and bacteria—chiefly gums, waxes, and other glue-like substances. These by-products cement the soil particles together, forming water-stable aggregates. The aggregate is then strong enough to hold together when wet—hence the term "water-stable."

USDA soil microbiologist Sara Wright named the glue that holds aggregates together "glomalin" after the Glomales group of common root-dwelling fungi. These fungi secrete a gooey protein known as glomalin through their hair-like filaments, or hyphae. When Wright measured glomalin in soil aggregates she found levels as high as 2% of their total weight in eastern U.S. soils.

Soil aggregates from the West and Midwest had lower levels of glomalin. She found that tillage tends to lower glomalin levels. Glomalin levels and aggregation were higher in no-till corn plots than in tilled plots. Wright has a brochure describing glomalin and how it benefits soil, entitled *Glomalin, a Manageable Soil Glue.* To order this brochure see the Additional Resources section of this publication.

A well-aggregated soil allows for increased water entry, increased air flow, and increased water-holding capacity. Plant roots occupy a larger volume of well-aggregated soil, high in organic matter, as compared to a finely pulverized and dispersed soil, low in organic matter. Roots, earthworms, and soil arthropods can pass more easily through a well-aggregated soil.

Aggregated soils also prevent crusting of the soil surface. Finally, well-aggregated soils are more erosion resistant, because aggregates are much heavier than their particle components. For a good example of the effect of organic matter additions on aggregation, as shown by subsequent increase in water entry into the soil.

Table : Water entry into the soil after 1 hour

Manure Rate (tons/acre)	*Inches of water*
0	1.2
8	1.9
16	2.7

Boyle et al.

The opposite of aggregation is dispersion. In a dispersed soil, each individual soil particle is free to blow away with the wind or wash away with overland flow of water.

Clay soils with poor aggregation tend to be sticky when wet, and cloddy when dry. If the clay particles in these soils can be aggregated together, better aeration and water infiltration will result. Sandy soils can benefit from aggregation by having a small amount of dispersed clay that tends to stick between the sand particles and slow the downward movement of water.

Crusting is a common problem on soils that are poorly aggregated. Crusting results chiefly from the impact of falling raindrops. Rainfall causes clay particles on the soil surface to disperse and clog the pores immediately beneath the surface. Following drying, a sealed soil surface results in which most of the pore space has been drastically reduced due to clogging from dispersed clay particles. Subsequent rainfall is much more likely to run off than to flow into the soil.

Since raindrops start crusting, any management practices that protect the soil from their impact will decrease crusting and increase water flow into the soil. Mulches and cover crops serve this purpose well, as do no-till practices, which allow the accumulation of surface residue. Also, a well-aggregated soil will resist crusting because the water-stable aggregates are less likely to break apart when a raindrop hits them.

Long-term grass production produces the best-aggregated soils. A grass sod extends a mass of fine roots throughout the topsoil, contributing to the physical processes that help form aggregates. Roots continually remove water from soil microsites, providing local wetting and drying effects that promote aggregation. Fine root hairs also bind soil aggregates together.

Roots also produce food for soil microorganisms and earthworms, which in turn generate compounds that bind soil particles into water-stable aggregates. In addition, perennial grass sods provide protection from raindrops and erosion. Thus, a perennial cover creates a combination of conditions optimal for the creation and maintenance of well-aggregated soil.

Conversely, cropping sequences that involve annual plants and extensive cultivation provide less vegetative cover and organic matter, and usually result in a rapid decline in soil aggregation. For more information on aggregation, see the soil quality information sheet entitled Aggregate Stability at the Soil Quality Institute's home page. From there, click on Soil Quality Information Sheets, then click on Aggregate Stability.

The best-aggregated soils are those that have been in long-term grass production.

Farming practices can be geared to conserve and promote soil aggregation. Because the binding substances are themselves susceptible to microbial degradation, organic matter needs to be replenished to maintain microbial populations and overall aggregated soil status. Practices should conserve aggregates once they are formed, by minimizing factors that degrade and destroy aggregation. Some factors that destroy or degrade soil aggregates are:

- bare soil surface exposed to the impact of raindrops
- removal of organic matter through crop production and harvest without return of organic matter to the soil
- excessive tillage
- working the soil when it is too wet or too dry
- use of anhydrous ammonia, which speeds up decomposition of organic matter
- excess nitrogen fertilization
- allowing the build-up of excess sodium from irrigation or sodium-containing fertilizers.

Tillage, Organic Matter, and Plant Productivity

Several factors affect the level of organic matter that can be maintained in a soil. Among these are organic matter additions, moisture, temperature, tillage, nitrogen levels, cropping, and fertilization. The level of organic matter present in the soil is a direct function of how much organic material is being produced or added to the soil versus the rate of decomposition. Achieving this balance entails slowing the speed of organic matter decomposition, while increasing the supply of organic materials produced on site and/or added from off site.

Moisture and temperature also profoundly affect soil organic matter levels. High rainfall and temperature promote rapid plant growth, but these conditions are also favourable to rapid organic matter decomposition and loss. Low rainfall or low temperatures slow both plant growth and organic matter decomposition. The native Midwest prairie soils originally had a high amount of organic matter from the continuous growth and decomposition of perennial grasses, combined with a moderate temperature that did not allow for rapid decomposition of organic matter. Moist and hot tropical areas may appear lush because of rapid plant growth, but soils in these areas are low in nutrients. Rapid decomposition of organic matter returns nutrients back to the soil, where they are almost immediately taken up by rapidly growing plants.

Tillage can be beneficial or harmful to a biologically active soil, depending on what type of tillage is used and when it is done. Tillage affects both erosion rates and soil organic matter decomposition rates. Tillage can reduce the organic matter level in croplands below 1%, rendering them biologically dead. Clean tillage involving moldboard plowing and disking breaks down soil aggregates and leaves the soil prone to erosion from wind and water. The moldboard plow can bury crop residue and topsoil to a depth of 14 inches. At this depth, the oxygen level in the soil is so low that decomposition cannot proceed adequately. Surface-dwelling decomposer organisms suddenly find themselves suffocated and soon die. Crop residues that were originally on the surface but now have been turned under will putrefy in the oxygen-deprived zone. This rotting activity may give a putrid smell to the soil. Furthermore, the top few inches of the field are now often covered with subsoil having very little organic matter content and, therefore, limited ability to support productive crop growth.

The topsoil is where the biological activity happens—it's where the oxygen is. That's why a fence post rots off at the surface. In terms of organic matter, tillage is similar to opening the air vents on a wood-burning stove; adding organic matter is like adding wood to the stove. Ideally, organic matter decomposition should proceed as an efficient burn of the "wood" to release nutrients and carbohydrates to the soil organisms and create stable humus. Shallow tillage incorporates residue and speeds the decomposition of organic matter by adding oxygen that microbes need to become more active.

In cold climates with a long dormant season, light tillage of a heavy residue may be beneficial; in warmer climates it is hard enough to maintain organic matter levels without any tillage.

The plow lays the soil up on its side, increasing the surface area exposed to oxygen. The other three types of tillage are intermediate in their ability to foster organic matter decomposition. Oxygen is the key factor here. The moldboard plow increases the soil surface area, allowing more air into the soil and speeding the decomposition rate. With the moldboard plow, more than the entire organic matter contribution from the wheat straw is gone within only 19 days following tillage. Finally, the passage of heavy equipment increases compaction in the wheel tracks, and some tillage implements themselves compact the soil further, removing oxygen and increasing the chance that deeply buried residues will putrefy.

Tillage also reduces the rate of water entry into the soil by removal of ground cover and destruction of aggregates, resulting in compaction and crusting. Notice the direct relationship between tillage type, ground cover, and water infiltration. No-till has more than three times the water infiltration of the moldboard-plowed soil. Additionally, no-till fields will have higher aggregation from the organic matter decomposition on site. The surface mulch typical of no-till fields acts as a protective skin for the soil. This soil skin reduces the impact of raindrops and buffers the soil from temperature extremes as well as reducing water evaporation.

Table : Tillage effects on water infiltration and ground cover.

	Water Infiltration mm/minute	*Ground Cover Percent*
No-till	2.7	48
Chisel Plow	1.3	27
Moldboard Plow	0.8	12

From Boyle et al., 1989.

Both no-till and reduced-tillage systems provide benefits to the soil. The advantages of a no-till system include superior soil conservation, moisture conservation, reduced water runoff, long-term buildup of organic matter, and increased water infiltration. A soil managed without tillage relies on soil organisms to take over the job of plant residue incorporation formerly done by tillage. On the down side, no-till can foster a reliance on herbicides to control weeds and can lead to soil compaction from the traffic of heavy equipment.

Pioneering development work on chemical-free no-till farming is proceeding at several research stations and farms in the eastern U.S. Pennsylvania farmer Steve Groff has been farming no-till with minimal or no herbicides for several years. Groff grows cover crops extensively in his fields, rolling them down in the spring using a 10-foot rolling stalk chopper. This rolling chopper kills the rye or vetch cover crop and creates a nice no-till mulch into which he plants a variety of vegetable and grain crops. After several years of no-till production, his soils are mellow and easy to plant into. Groff farms 175 acres of vegetables, alfalfa, and grain crops on his Cedar Meadow Farm. Learn more about his operation in the Farmer Profiles section of this publication, by visiting his Web site, or by ordering his video.

Other conservation tillage systems include ridge tillage, minimum tillage, zone tillage, and reduced tillage, each possessing some of the advantages of both conventional till and no-till. These systems represent intermediate tillage systems, allowing more flexibility than either a no-till or conventional till system might. They are more beneficial to soil organisms than a conventional clean-tillage system of moldboard plowing and disking.

Adding manure and compost is a recognized means for improving soil organic matter and humus levels. In their absence, perennial grass is the only crop that can regenerate and increase soil humus. Cool-season grasses build soil organic matter faster than warm-season grasses because they are growing much longer during a given year. hen the soil is warm enough for soil organisms to decompose organic matter, cool-season grass is growing. While growing, it is producing organic matter and cycling minerals from the decomposing organic matter in the soil. In other words, there is a net gain of organic matter because the cool-season grass is producing organic matter faster than it is being used up. With warm-season grasses, organic matter production during the growing season can be slowed during the long dormant season from fall through early spring. During the beginning and end of this dormant period, the soil is still biologically active, yet no grass growth is proceeding. Some net accumulation of organic matter can occur under warm-season grasses, however. In a Texas study, switchgrass (a warm-season grass) grown for four years increased soil carbon content from 1.1% to 1.5% in the top 12 inches of soil. In hot and moist regions, a cropping rotation that includes several years of pasture will be most beneficial.

Effect of Nitrogen on Organic Matter

Excessive nitrogen applications stimulate increased microbial activity, which in turn speeds organic matter decomposition. The extra nitrogen narrows the ratio of carbon to nitrogen in the soil. Native or uncultivated soils have approximately 12 parts of carbon to each part of nitrogen, or a C:N ratio of 12:1. At this ratio, populations of decay bacteria are kept at a stable level, since additional growth in their population is limited by a lack of nitrogen. When large amounts of inorganic nitrogen are added, the C:N ratio is reduced, which allows the populations of decay organisms to explode as they decompose more organic matter with the now abundant nitrogen. While soil bacteria can efficiently use moderate applications of inorganic nitrogen accompanied by organic amendments (carbon), excess nitrogen results in decomposition of existing organic matter at a rapid rate. Eventually, soil carbon content may be reduced to a level where the bacterial populations are on a starvation diet. With little carbon available, bacterial populations shrink, and less of the free soil nitrogen is absorbed. Thereafter, applied nitrogen, rather than being cycled through microbial organisms and re-released to plants slowly over time, becomes subject to leaching. This can greatly reduce the efficiency of fertilization and lead to environmental problems.

Excessive nitrogen stimulates increased microbial activity, which in turn speeds organic matter decomposition.

To minimize the fast decomposition of soil organic matter, carbon should be added with nitrogen. Typical carbon sources—such as green manures, animal manure, and compost—serve this purpose well.

Amendments containing too high a carbon to nitrogen ratio (25:1 or more) can tip the balance the other way, resulting in nitrogen being tied up in an unavailable form. Soil organisms consume all the nitrogen in an effort to decompose the abundant carbon; tied up in the soil organisms, nitrogen remains unavailable for plant uptake. As soon as a soil microorganism dies and decomposes, its nitrogen is consumed by another soil organism, until the balance between carbon and nitrogen is achieved again.

Fertilizer Amendments and Biologically Active Soils

What are the soil mineral conditions that foster biologically active soils? Drawing from the work of Dr. William Albrecht (1888 to 1974), agronomist at the University of Missouri, we learn that balance is the key. Albrecht advocated bringing soil nutrients into a balance so that none were in excess or deficient. Albrecht's theory (also called base-saturation theory) is used to guide lime and fertilizer application by measuring and evaluating the ratios of positively charged nutrients (bases) held in the soil. Positively charged bases include calcium, magnesium, potassium, sodium, ammonium nitrogen, and several trace minerals. When optimum ratios of bases exist, the soil is believed to support high biological activity, have optimal physical properties (water intake and aggregation), and become resistant to leaching. Plants growing on such a soil are also balanced in mineral levels and are considered to be nutritious to humans and animals alike. Base saturation percentages that Albrecht's research showed to be optimal for the growth of most crops are:

Calcium	60-70%
Magnesium	10-20%
Potassium	2-5%
Sodium	0.5-3%
Other bases	5%

According to Albrecht, fertilizer and lime applications should be made at rates that will bring soil mineral percentages into this ideal range. This approach will shift the soil pH automatically into a desirable range without creating nutrient imbalances. The base saturation theory also takes into account the effect one nutrient may have on another and avoids undesirable interactions. For example, phosphorus is known to tie up zinc.

The Albrecht system of soil evaluation contrasts with the approach used by many state laboratories, often called the "sufficiency method." Sufficiency theory places little to no value on nutrient ratios, and lime recommendations are typically based on pH measurements alone. While in many circumstances base saturation and sufficiency methods will produce identical soil recommendations and similar results, significant differences can occur on a number of soils. For example, suppose we tested a cornfield and found a soil pH of 5.5 and base saturation for magnesium at 20% and calcium at 40%. Base saturation theory would call for liming with a high-calcium lime to raise the percent base saturation of calcium; the pH would rise accordingly. Sufficiency theory would not specify high-calcium lime and the grower might choose instead a high-magnesium dolomite lime that would raise the pH but worsen the balance of nutrients in the soil. Another way to look at these two theories is that the base saturation theory does not concern itself with pH to any great extent, but rather with the proportional amounts of bases. The pH will be correct when the levels of bases are correct.

Albrecht's ideas have found their way onto large numbers of American farms and into the programs of several agricultural consulting companies. Neal Kinsey, a soil fertility consultant in Charleston, Missouri, is a major proponent of the Albrecht approach. Kinsey was a student under Albrecht and is one of the leading authorities on the base-saturation method. He teaches a short course on the Albrecht system and provides a soil analysis service. His book, *Hands On Agronomy*, is widely recognized as a highly practical guide to the Albrecht system. ATTRA can provide more information on Albrecht Fertility Management Systems.

Several firms—many providing backup fertilizer and amendment products—offer a biological-farming program based on the Albrecht theory. Typically these firms offer broad-based soil analysis and recommend balanced fertilizer materials considered friendly to soil organisms. They avoid the use of some common fertilizers and amendments such as dolomite lime, potassium chloride, anhydrous ammonia, and oxide forms of trace elements because they are considered harmful to soil life. The publication *How to Get Started in Biological Farming* presents such a program. For names of companies offering consulting and products, order the ATTRA publications Alternative Soil Testing Laboratories and Sources for Organic Fertilizers and Amendments. Both of these are also available on the ATTRA Web site.

Conventional Fertilizers

Commercial fertilizer can be a valuable resource to farmers in transition to a more sustainable system and can help meet nutrient needs during times of high crop nutrient demand or when weather conditions result in slow nutrient release from organic resources. Commercial fertilizers have the advantage of supplying plants with immediately available forms of nutrients. They are often less expensive and less bulky to apply than many natural fertilizers.

Not all conventional fertilizers are alike. Many appear harmless to soil livestock, but some are not. Anhydrous ammonia contains approximately 82% nitrogen and is applied subsurface as a gas. Anhydrous speeds the decomposition of organic matter in the soil, leaving the soil more compact as a result. The addition of anhydrous causes increased acidity in the soil, requiring 148 pounds of lime to neutralize 100 pounds of anhydrous ammonia, or 1.8 pounds of lime for every pound of nitrogen contained in the anhydrous. Anhydrous ammonia initially kills many soil microorganisms in the application zone. Bacteria and actinomycetes recover within one to two weeks to levels higher than those prior to treatment. Soil fungi, however, may take seven weeks to recover.

During the recovery time, bacteria are stimulated to grow more, and decompose more organic matter, by the high soil nitrogen content. As a result, their numbers increase after anhydrous applications, then decline as available soil organic matter is depleted. Farmers commonly report that the long-term use of synthetic fertilizers, especially anhydrous ammonia, leads to soil compaction and poor tilth. When bacterial populations and soil organic matter decrease, aggregation declines, because existing glues that stick soil particles together are degraded, and no other glues are being produced.

Potassium chloride (KCl) (0-0-60 and 0-0-50), also known as muriate of potash, contains approximately 50 to 60% potassium and 47.5% chloride. Muriate of potash is made by refining potassium chloride ore, which is a mixture of potassium and sodium salts and clay from the brines of dying lakes and seas. The potential harmful effects from KCl can be surmised from the salt concentration of the material. Additionally, some plants such as tobacco, potatoes, peaches, and some legumes are especially sensitive to chloride. High rates of KCl must be avoided on such crops. Potassium sulfate, potassium nitrate, sul-po-mag, or organic sources of potassium may be considered as alternatives to KCl for fertilization.

Sodium nitrate, also known as Chilean nitrate or nitrate of soda, is another high-salt fertilizer. Because of the relatively low nitrogen content of sodium nitrate, a high amount of sodium is added to the soil when normal applications of nitrogen are made with this material. The concern is that excessive sodium acts as a dispersant of soil particles, degrading aggregation.

Table : Salt index for various fertilizers.

Material	*Salt Index*	*Salt index per unit of plant food*
Sodium chloride	153	2.90
Potassium chloride	116	1.90
Ammonium nitrate	105	3.00
Sodium nitrate	100	6.10
Urea	75	1.60
Potassium nitrate	74	1.60
Ammonium sulfate	69	3.30
Calcium nitrate	53	4.40
Anhydrous ammonia	47	0.06
Sulfate-potash-magnesia	43	2.00
Di-ammonium phosphate	34	1.60
Monammonium phosphate	30	2.50
Gypsum	8	0.03
Calcium carbonate	5	0.01

Top$oil—Your Farm'$ Capital

Protecting soil from erosion is the first step toward a sustainable agriculture.

Topsoil is the capital reserve of every farm. Ever since mankind started agriculture, erosion of topsoil has been the single largest threat to a soil's productivity—and, consequently, to farm profitability. This is still true today. In the U.S., the average acre of cropland is eroding at a rate of 7 tons per year. To sustain agriculture means to sustain soil resources, because that's the source of a farmer's livelihood.

The major productivity costs to the farm associated with soil erosion come from the replacement of lost nutrients and reduced water holding ability, accounting for 50 to 75% of productivity loss.

Soil that is removed by erosion typically contains about three times more nutrients than the soil left behind and is 1.5 to 5 times richer in organic matter. This organic matter loss not only results in reduced water holding capacity and degraded soil aggregation, but also loss of plant nutrients, which must then be replaced with nutrient amendments.

Five tons of topsoil (the so-called tolerance level) can easily contain 100 pounds of nitrogen, 60 pounds of phosphate, 45 pounds of potash, 2 pounds of calcium, 10 pounds of magnesium, and 8 pounds of sulfur. Moderate, and severe erosion on organic matter, soil phosphorus level, and plant-available water on a silt loam soil in Indiana.

Table : Effect of erosion on organic matter phosphorus and plant-available water.

Erosion Level	Organic matter	Phosphorus	Plant-available water
	(%)	(lbs./ac)	(%)
Slight	3.0	62	7.4
Moderate	2.5	61	6.2
Severe	1.9	40	3.6

From Schertz et al., 1984.

When erosion by water and wind occurs at a rate of 7.6 tons/acre/year it costs $40 per acre each year to replace the lost nutrients as fertilizer and around $17/acre/year to pump well irrigation water to replace the soil water holding capacity of that lost soil. The total cost of soil and water lost annually from U.S. cropland amounts to an on-site productivity loss of approximately $27 billion each year.

Water erosion gets started when falling rainwater collides with bare ground and detaches soil particles from the parent soil body. After enough water builds up on the soil surface, following detachment, overland water flow transports suspended soil down-slope. Suspended soil in the runoff water abrades and detaches additional soil particles as the water travels overland. Preventing detachment is the most effective point of erosion control because it keeps the soil in place. Other erosion control practices seek to slow soil particle transport and cause soil to be deposited before it reaches streams. These methods are less effective at protecting the quality of soil within the field. Commonly implemented practices to slow soil transport include terraces and diversions. Terraces, diversions, and many other erosion "control" practices are largely unnecessary if the ground stays covered year-round. For erosion prevention, a high percentage of ground cover is a good indicator of success, while bare ground is an "early warning" indicator for a high risk of erosion. Muddy runoff water and gullies are "too-late" indicators. The soil has already eroded by the time it shows up as muddy water, and it's too late to save soil already suspended in the water.

Protecting the soil from erosion is *the* first step toward a sustainable agriculture. Since water erosion is initiated by raindrop impact on bare soil, any management practice that protects the soil from raindrop impact will decrease erosion and increase water entry into the soil. Mulches, cover crops, and crop residues serve this purpose well. Additionally, well-aggregated soils resist crusting because water-stable aggregates are less likely to break apart when the raindrop hits them. Adequate organic matter with high soil biological activity leads to high soil aggregation.

Many studies have shown that cropping systems that maintain a soil-protecting plant canopy or residue cover have the least soil erosion. This is universally true. Long-term cropping studies begun in 1888 at the University of Missouri provide dramatic evidence of this. Gantzer and colleagues examined the effects of a century of cropping on soil erosion. They compared depth of topsoil remaining after 100 years of cropping. As the table shows, the cropping system that maintained the highest amount of permanent ground cover (timothy grass) had the greatest amount of topsoil left.

Table : Topsoil depth remaining after 100 years of different cropping practices.

Crop Sequence	*Inches of topsoil remaining*
Continuous Corn	7.7
6-year rotation*	12.2
Continuous timothy grass	17.4

*Corn, oats, wheat, clover, timothy, timothyFrom: Gantzer et al.

The researchers commented that subsoil had been mixed with topsoil in the continuous corn plots from plowing, making the real topsoil depth less than was apparent. In reality, all the topsoil was lost from the continuous corn plots in only 100 years. The rotation lost about half the topsoil over 100 years. How can we feed future generations with this type of farming practice? In a study of many different soil types in each of the major climatic zones of the U.S., researchers showed dramatic differences in soil erosion when comparing row crops to perennial sods. Row crops consisted of cotton or corn, and sod crops were bluegrass or bermuda grass. On average, the row crops eroded more than 50 times more soil than did the perennial sod crops. The two primary influencing factors are ground cover and tillage.

Table : Effect of cropping on soil erosion rates.

Soil Type	*Location (State)*	*Slope (%)*	*Row crop soil loss (Tons/ac)*	*Sod soil loss (Tons/ac)*
Silt loam	Iowa	9.0	38	.02
Loam	Missouri	8.0	51	.16
Silt loam	Ohio	12.0	99	.02
Fine sandy loam	Oklahoma	7.7	19	.02
Clay loam	N. Carolina	10.0	31	.31
Fine sandy loam	Texas	8.7	24	.08
Clay	Texas	4.0	21	.02
Silt loam	Wisconsin	16.0	111	.10
Aberage	Average	9.4	49	.09

Adapted from Shiflet and Darby, 1985.

So, how long do fields have before the topsoil is gone? This depends on where in the country the field is located. Some soils naturally have very thick topsoil, while other soils have thin topsoil over rock or gravel. Roughly 8 tons/acre/year of soil-erosion loss amounts to the thickness of a dime spread over an acre. Twenty dimes stack up to 1-inch high. So a landscape with an 8-ton erosion rate would lose an inch of topsoil about every 20 years. On a soil with a thick topsoil, this amount is barely detectable within a person's lifetime and may not be noticed. Soils with naturally thin topsoils or topsoils that have been previously eroded can be transformed from productive to degraded land within a generation.

Forward-thinking researcher Wes Jackson, of the Land Institute, waxes eloquent about how tillage has become engrained in human culture since we first began farming. Beating our swords into plowshares surely embodies the triumph of good over evil. Someone who creates something new is said to have "plowed new ground." "Yet the plowshare may well have destroyed more options for future generations than the sword."

Tillage for the production of annual crops is the major problem in agriculture, causing soil erosion and the loss of soil quality. Any agricultural practice that creates and maintains bare ground is inherently less sustainable than practices that keep the ground covered throughout the year. Wes Jackson has spent much of his career developing perennial grain crops and cropping systems that mimic the natural prairie. Perennial grain crops do not require tillage to establish year after year, and the ground is left covered. Ultimately, this is the future of grain production and truly represents a new vision for how we produce food. The greatest research need in agriculture today is breeding work to develop perennial crops that will replace annual crops requiring tillage. Farming practices using annual crops in ways that mimic perennial systems, such as no-till and cover crops, are our best alternative until perennial systems are developed.

Summary of Part I

Soil management involves stewardship of the soil livestock herd. The primary factors affecting organic matter content, build-up, and decomposition rate in soils are oxygen content, nitrogen content, moisture content, temperature, and the addition and removal of organic materials. All these factors work together all the time. Any one can limit the others. These are the factors that affect the health and reproductive rate of organic matter decomposer organisms. Managers need to be aware of these factors when making decisions about their soils. Let's take them one at a time.

Increasing oxygen speeds decomposition of organic matter. Tillage is the primary way extra oxygen enters the soil. Texture also plays a role, with sandy soils having more aeration than heavy clay soils. Nitrogen content is influenced by fertilizer additions. Excess nitrogen, without the addition of carbon, speeds the decomposition of organic matter. Moisture content affects decomposition rates. Soil microbial populations are most active over cycles of wetting and drying. Their populations increase following wetting, as the soil dries out. After the soil becomes dry, their activity diminishes. Just like humans, soil organisms are profoundly affected by temperature. Their activity is highest within a band of optimum temperature, above and below which their activity is diminished.

Adding organic matter provides more food for microbes. To achieve an increase of soil organic matter, additions must be higher than removals. Over a given year, under average conditions, 60 to 70 percent of the carbon contained in organic residues added to soil is lost as carbon dioxide. Five to ten percent is assimilated into the organisms that decomposed the organic residues, and the rest becomes 'new' humus. It takes decades for new humus to develop into stable humus, which imparts the nutrient-holding characteristics humus is known for. The end result of adding a ton of residue would be 400 to 700 pounds of new humus. One percent organic matter weighs 20,000 pounds per acre. A 7-inch depth of topsoil over an acre weighs 2 million pounds. Building organic matter is a slow process.

It is more feasible to stabilize and maintain the humus present, before it is lost, than to try to rebuild it. The value of humus is not fully realized until it is severely depleted. If your soils are high in humus now, work hard to preserve what you have. The formation of new humus is essential to maintaining old humus, and the decomposition of raw organic matter has many benefits of its own. Increased aeration caused by tillage coupled with the absence of organic carbon in fertilizer materials has caused more than a 50% decline in native humus levels on many U.S. farms.

Appropriate mineral nutrition needs to be present for soil organisms and plants to prosper. Adequate levels of calcium, magnesium, potassium, phosphorus, sodium, and the trace elements should be present, but not in excess. The base saturation theory of soil management helps guide decision-making toward achieving optimum levels of these nutrients in the soil. Several books have been written on balancing soil mineral levels, and several consulting firms provide soil analysis and fertility recommendation services based on this theory.

Commercial fertilizers have their place in sustainable agriculture. Some appear harmless to soil livestock and provide nutrients at times of high nutrient demand from crops. Anhydrous ammonia and potassium chloride cause problems, however. As noted above, anhydrous kills soil organisms in the injection zone. Bacteria and actinomycetes recover within a few weeks, but fungi take longer. The increase in bacteria, fed by highly available nitrogen from the anhydrous, speeds the decomposition of organic matter. Potassium chloride has a high salt index, and some plants and soil organisms are sensitive to chloride.

Topsoil is the farmer's capital. Sustaining agriculture means sustaining the soil. Maintaining ground cover in the form of cover crops, mulch, or crop residue for as much of the annual season as possible achieves the goal of sustaining the soil resource. Any time the soil is tilled and left bare it is susceptible to erosion. Even small amounts of soil erosion are harmful over time. It is not easy to see the effects of erosion over a human lifetime; therefore, erosion may go unnoticed. Tillage for production of annual crops has created most of the erosion associated with agriculture. Perennial grain crops not requiring tillage provide a promising alternative for drastically improving the sustainability of future grain production.

Summary of Sustainable Soil Management Principles

- Soil livestock cycle nutrients and provide many other benefits.
- Organic matter is the food for the soil livestock herd.
- The soil should be covered to protect it from erosion and temperature extremes.
- Tillage speeds the decomposition of organic matter.
- Excess nitrogen speeds the decomposition of organic matter; insufficient nitrogen slows down organic matter decomposition and starves plants.
- Moldboard plowing speeds the decomposition of organic matter, destroys earthworm habitat, and increases erosion.
- To build soil organic matter, the production or addition of organic matter must exceed the decomposition of organic matter.
- Soil fertility levels need to be within acceptable ranges before a soil-building program is begun.

Part II. Management Steps to Improve Soil Quality

Assess Soil Health and Biological Activity on Your Farm

A basic soil audit is the first and sometimes the only monitoring tool used to assess changes in the soil. Unfortunately, the standard soil test done to determine nutrient levels (P, K, Ca, Mg, etc.) provides no information on soil biology and physical properties. Yet most of the farmer-recognized criteria for healthy soils include, or are created by, soil organisms and soil physical properties. A better appreciation of these biological and physical soil properties, and how they affect soil management and productivity, has resulted in the adoption of several new soil health assessment techniques, which are discussed below.

The USDA Soil Quality Test Kit

The USDA Soil Quality Institute provides a Soil Quality Test Kit Guide developed by Dr. John Doran and associates at the Agricultural Research Service's office in Lincoln, Nebraska. Designed for field use, the kit allows the measurement of water infiltration, water holding capacity, bulk density, pH, soil nitrate, salt concentration, aggregate stability, earthworm numbers, and soil respiration. Components necessary to build a kit include many items commonly available—such as pop bottles, flat-bladed knives, a garden trowel, and plastic wrap. Also necessary to do the tests is some equipment that is not as readily available, such as hypodermic needles, latex tubing, a soil thermometer, an electrical conductivity meter, filter paper, and an EC calibration standard.

A greatly simplified and quick soil quality assessment is available at the Soil Quality Institute's Web page as well, by clicking on "Getting to Know your Soil," near the bottom of the homepage. This simplified method involves digging a hole and making some observations. Here are a few of the procedures shown at this Web site: Dig a hole 4 to 6 inches below the last tillage depth and observe how hard the digging is. Inspect plant roots to see whether there is a lot of branching and fine root hairs or whether the roots are balled-up. A lack of fine root hairs indicates oxygen deprivation, while sideways growth indicates a hardpan. The process goes on to assess earthworms, soil smell, and aggregation. Another useful, hands-on procedure for assessing pasture soils is discussed in the ATTRA publication Assessing the Pasture Soil Resource.

Early Warning Monitoring for Croplands

A cropland monitoring guide has been published by the Centre for Holistic Management. The guide contains a set of soil health indicators that are measurable in the field. No fancy equipment is needed to make the assessments described in this monitoring guide. In fact, all the equipment is cheap and locally available for almost any farm. Simple measurements can help determine the health of croplands in terms of the effectiveness of the nutrient cycle and water cycle, and the diversity of some soil organisms.

Assessments of living organisms, aggregation, water infiltration, ground cover, and earthworms can be made using this guide. The monitoring guide is easy to read and understand and comes with a field sheet to record observations. It is available for $12 from the Savory Centre for Holistic Management.

Direct Assessment of Soil Health

Some quick ways to identify a healthy soil include feeling it and smelling it. Grab a handful and take a whiff. Does it have an earthy smell? Is it a loose, crumbly soil with some earthworms present? Dr. Ray Weil, soil scientist at the University of Maryland, describes how he would make a quick evaluation of a soil's health in just five minutes.

Look at the surface and see if it is crusted, which tells something about tillage practices used, organic matter, and structure. Push a soil probe down to 12 inches, lift out some soil and feel its texture. If a plow pan were present it would have been felt with the probe. Turn over a shovelful of soil to look for earthworms and smell for actinomycetes, which are microorganisms that help compost and stabilize decaying organic matter. Their activity leaves a fresh earthy smell in the soil.

Two other easy observations are to count the number of soil organisms in a square foot of surface crop residue and to pour a pint of water on the soil and record the time it takes to sink in. Comparisons can be made using these simple observations, along with Ray Weil's evaluation above, to determine how farm practices affect soil quality. Some of the soil quality assessment systems discussed above use these and other observations and provide record keeping sheets to record your observations.

A Simple Erosion Demonstration

This simple procedure demonstrates the value of ground cover. Tape a white piece of paper near the end of a three-foot-long stick. Hold the stick in one hand so as to have the paper end within one inch of a bare soil surface Now pour a pint of water onto the bare soil within two to three inches of the white paper and observe the soil accumulation on the white paper. Tape another piece of white paper to the stick and repeat the operation, this time over soil with 100% ground cover, and observe the accumulation of soil on the paper. Compare the two pieces of paper. This simple test shows how effective ground cover can be at preventing soil particles from detaching from the soil surface.

Use Tools and Techniques to Build Soil

Can a cover crop be worked into your rotation? How about a high-residue crop or perennial sod? Are there economical sources of organic materials or manure in your area? Are there ways to reduce tillage and nitrogen fertilizer? Where feasible, bulky organic amendments may be added to supply both organic matter and plant nutrients. It is particularly useful to account for nutrients when organic fertilizers and amendments are used. Start with a soil test and a nutrient analysis of the material you are applying. Knowing the levels of nutrients needed by the crop guides the amount of amendments applied and can lead to significant reductions in fertilizer cost. The nutrient composition of organic materials can vary, which is all the more reason to determine the amount you have with appropriate testing. In addition to containing the major plant nutrients, organic fertilizers can supply many essential micronutrients. Proper calibration of the spreading equipment is important to ensure accurate application rates.

Animal Manure

Manure is an excellent soil amendment, providing both organic matter and nutrients. The amount of organic matter and nitrogen in animal manure depends on the feed the animals consumed, type of bedding used (if any), and whether the manure is applied as a solid or liquid. Typical rates for dairy manure would be 10 to 30 tons per acre or 4,000 to 11,000 gallons of liquid for corn. At these rates the crop would get between 50 and 150 pounds of available nitrogen per acre. Additionally, lots of carbon would be added to the soil, resulting in no loss of soil organic matter. Residues from crops grown with this manure application and left on the soil would also contribute organic matter.

However, a common problem with using manure as a nutrient source is that application rates are usually based on the nitrogen needs of the crop. Because some manures have about as much phosphorus as they do nitrogen, this often leads to a buildup of soil phosphorus. A classic example is chicken litter applied to crops that require high nitrogen levels, such as pasture grasses and corn. Broiler litter, for example, contains approximately 50 pounds of nitrogen and phosphorus and about 40 pounds of potassium per ton.

Since an established fescue pasture needs twice as much nitrogen as it does phosphorus, a common fertilizer application would be about 50 pounds of nitrogen and 30 pounds of phosphorus per acre. If a ton of poultry litter were applied to supply the nitrogen needs of the fescue, an over-application of phosphorus would result, because the litter has about the same levels of nitrogen and phosphorus. Several years of litter application to meet nitrogen needs can build up soil phosphorus to excessive levels. One easy answer to this dilemma is to adjust the manure rate to meet the phosphorus needs of the crop and to supply the additional nitrogen with fertilizer or a legume cover crop. On some farms this may mean that more manure is being produced than can be safely used on the farm. In this case, farmers may need to find a way to process and sell (or barter) this excess manure to get it off the farm.

Compost

Composting farm manure and other organic materials is an excellent way to stabilize their nutrient content. Composted manure is also easier to handle, less bulky, and better smelling than raw manure. A significant portion of raw-manure nutrients are in unstable, soluble forms. Such unstable forms are more likely to run off if surface-applied, or to leach if tilled into the soil. Compost is not as good a source of readily available plant nutrients as raw manure. But compost releases its nutrients slowly, thereby minimizing losses. Quality compost contains more humus than its raw components because primary decomposition has occurred during the composting process.

However, it does not contribute as much of the sticky gums and waxes that aggregate soil particles together as does raw manure, because these substances are also released during the primary decomposition phase. Unlike manure, compost can be used at almost any rate without burning plants. In fact, some greenhouse potting mixes contain 20 to 30% compost. Compost (like manure) should be analysed by a laboratory to determine the nutrient value of a particular batch and to ensure that it is being used effectively to produce healthy crops and soil, and not excessively so that it contributes to water pollution.

Composting also reduces the bulk of raw organic materials—especially manures, which often have a high moisture content. However, while less bulky and easier to handle, composts can be expensive to buy. On-farm composting cuts costs dramatically, compared with buying compost. For more comprehensive information on composting at the farm or the municipal level, see the ATTRA publication Farm-Scale Composting Resource List.

Cover Crops and Green Manures

Many types of plants can be grown as cover crops. Some of the more common ones include rye, buckwheat, hairy vetch, crimson clover, subterranean clover, red clover, sweet clover, cowpeas, millet, and forage sorghums. Each of these plants has some advantages over the others and differs in its area of adaptability. Cover crops can maintain or increase soil organic matter if they are allowed to grow long enough to produce high herbage. All too often, people get in a hurry and take out a good cover crop just a week or two before it has reached its full potential. Hairy vetch or crimson clover can yield up to 2.5 tons per acre if allowed to go to 25% bloom stage. A mixture of rye and hairy vetch can produce even more.

In addition to organic matter benefits, legume cover crops provide considerable nitrogen for crops that follow them. Consequently, the nitrogen rate can be reduced following a productive legume cover crop taken out at the correct time. For example, corn grown following two tons of hairy vetch should produce high yields of grain with only half of the normal nitrogen application.

When small grains such as rye are used as cover crops and allowed to reach the flowering stage, additional nitrogen may be required to help offset the nitrogen tie-up caused by the high carbon addition of the rye residue. The same would be true of any high-carbon amendment, such as sawdust or wheat straw. Cover crops also suppress weeds, help break pest cycles, and through their pollen and nectar provide food sources for beneficial insects and honeybees. They can also cycle other soil nutrients, making them available to subsequent crops as the green manure decomposes. For more information on cover crops, see ATTRA's Overview of Cover Crops and Green Manures. This publication is comprehensive and provides many references to other available information on growing cover crops.

Humates

Humates and humic acid derivatives are a diverse family of products, generally obtained from various forms of oxidized coal. Coal-derived humus is essentially the same as humus extracts from soil, but there has been a reluctance in some circles to accept it as a worthwhile soil additive. In part, this stems from a belief that only humus derived from recently decayed organic matter is beneficial. It is also true that the production and recycling of organic matter in the soil cannot be replaced by coal-derived humus.

However, while sugars, gums, waxes and similar materials derived from fresh organic-matter decay play a vital role in both soil microbiology and structure, they are not humus. Only a small portion of the organic matter added to the soil will ever be converted to humus. Most will return to the atmosphere as carbon dioxide as it decays. Some studies have shown positive effects of humates, while other studies have shown no such effects. Generally, the consensus is that they work well in soils with low organic matter. In small amounts they do not produce positive results on soils already high in organic matter; at high rates they may tie up soil nutrients.

There are many humate products on the market. They are not all the same. Humate products should be evaluated in a small test plot for cost effectiveness before using them on a large scale. Salespeople sometimes make exaggerated claims for their products. ATTRA can provide more information on humates upon request.

Reduced Tillage

While tillage has become common to many production systems, its effects on the soil can be counter-productive. Tillage smoothes the soil surface and destroys natural soil aggregations and earthworm channels. Porosity and water infiltration decrease following most tillage operations. Plow pans may develop in many situations, particularly if soils are plowed with heavy equipment or when the soil is wet. Tilled soils have much higher erosion rates than soils left covered with crop residue.

Because of all the problems associated with conventional tillage operations, acreage under reduced tillage systems is increasing in America. Any tillage system that leaves in excess of 30% surface residue is considered a "conservation tillage" system by USDA. Conservation tillage includes no-till, zero-till, ridge-till, zone-till, and some variations of chisel plowing and disking. These conservation till strategies and techniques allow for establishing crops into the previous crop's residues, which are purposely left on the soil surface. The principal benefits of conservation tillage are reduced soil erosion and improved water retention in the soil, resulting in more drought resistance. Additional benefits that many conservation tillage systems provide include reduced fuel consumption, flexibility in planting and harvesting, reduced labour requirements, and improved soil tilth. Two of the most common conservation tillage systems are ridge tillage and no-till.

Ridge tillage is a form of conservation tillage that uses specialized planters and cultivators to maintain permanent ridges on which row crops are grown. After harvest, crop residue is left until planting time. To plant the next crop, the planter places the seed in the top of the ridge after pushing residue out of the way and slicing off the surface of the ridge top. Ridges are re-formed during the last cultivation of the crop.

Often, a band of herbicide is applied to the ridge top during planting. With banded herbicide applications, two cultivations are generally used: one to loosen the soil and another to create the ridge later in the season. No cultivation may be necessary if the herbicide is applied by broadcasting rather than banding. Because ridge tillage relies on cultivation to control weeds and reform ridges, this system allows farmers to further reduce their dependence on herbicides, compared with either conventional till or strict no-till systems.

Maintenance of the ridges is key to successful ridge tillage systems. The equipment must accurately reshape the ridge, clean away crop residue, plant in the ridge centre, and leave a viable seedbed. Not only does the ridge-tillage cultivator remove weeds, it also builds up the ridge. Harvesting in ridged fields may require tall, narrow dual wheels fitted to the combine. This modification permits the combine to straddle several rows, leaving the ridges undisturbed. Similarly, grain trucks and wagons cannot be driven randomly through the field. Maintenance of the ridge becomes a consideration for each process.

Conventional no-till methods have been criticized for a heavy reliance on chemical herbicides for weed control. Additionally, no-till farming requires careful management and expensive machinery for some applications. In many cases, the spring temperature of untilled soil is lower than that of tilled soil. This lower temperature can slow germination of early-planted corn or delay planting dates. Also, increased insect and rodent pest problems have been reported. On the positive side, no-till methods offer excellent soil erosion prevention and decreased trips across the field. On well-drained soils that warm adequately in the spring, no-till has provided the same or better yields than conventional till.

A recent equipment introduction into the no-till arena is the so-called "no-till cultivator." These cultivators permit cultivation in heavy residue and provide a non-chemical option to post-emergent herbicide applications. Farmers have the option to band herbicide in the row and use the no-till cultivator to clean the middles as a way to reduce herbicide use. ATTRA can provide a number of resource contacts on cultural methods, equipment, and management for conservation-till cropping systems.

Minimize Synthetic Nitrogen Use

If at all possible, add carbon with nitrogen sources. Animal manure is a good way to add both carbon and nitrogen. Growing legumes as a green manure or rotation crop is another way. When using nitrogen fertilizer, try to do it at a time when a heavy crop residue is going onto the soil, too. For example, a rotation of corn, beans, and wheat would do well with nitrogen added after the corn residue was rolled down or lightly tilled in. Spring-planted soybeans would require no nitrogen.

A small amount of nitrogen could be applied in the fall for the wheat. Following the wheat crop, a legume winter-annual cover crop could be planted. In the spring, when the cover crop is taken out, nitrogen rates for the corn would be reduced to account for the nitrogen in the legume. Avoid continual hay crops accompanied by high nitrogen fertilization. The continual removal of hay accompanied by high nitrogen speeds the decomposition of soil organic matter. Heavy fertilization of silage crops, where all the crop residue is removed (especially when accompanied by tillage), speeds soil decline and organic matter depletion.

Continue to Monitor for Indicators of Success or Failure

As you experiment with new practices and amendments, continue to monitor the soil for changes using some of the tools discussed above in Assess Soil Health and Biological Activity. Several of these monitoring guides have sheets you can use in the field to record data and use for future comparison after changes are made to the farming practices.

Review the principles of sustainable soil management and find ways to apply them in your operation. If the thought of pulling everything together seems overwhelming, start with only one or two new practices and build on them. Seek additional motivation by reading the next section on people who have successfully built their soils.

Part III. Examples of Successful Soil Builders (Farmer Profiles)

Steve Groff

Steve and his family produce vegetables, alfalfa, and grain crops on 175 acres in Lancaster County, Pennsylvania. When Steve took over operation of the family farm 15 years ago, his number one concern was eliminating soil erosion. Consequently, he began using cover crops extensively in his fields. In order to transform his green cover crop into no-till mulch, Steve uses a 10-foot Buffalo rolling stalk chopper. Under the hitch-mounted frame, the stalk chopper has two sets of rollers running in tandem. These rollers can be adjusted for light or aggressive action and set for continuous coverage. Steve says the machine can be run up to eight miles an hour and does a good job of killing the cover crop and pushing it right down on the soil. It can also be used to flatten down other crop residues after harvest. Steve improved his chopper by adding independent linkages and springs to each roller. This modification makes each unit more flexible and allows continuous use over uneven terrain. Other farmers report similar results using a disk harrow with the gangs set to run straight or at a slight angle. Following his cover crop chopping, Steve transplants vegetable seedlings into the killed mulch; sweet corn and snap beans are direct-seeded. Since conversion to a cover crop mulch system, his soils are protected from erosion and have become much mellower. For more information on his system, order Steve's videos listed under Additional Resources, or visit his Web page. At Steve's Web site you can see photos of his cover crop roller and no-till transplanter in action, as well as test-plot results comparing flail mowing, rolling, and herbicide killing of cover crops.

Bob Willett

Bob started no-tilling 20 years ago on his corn and soybean farm in Pride, Kentucky. He not only reduced his machinery costs by switching to no-till but also made gains in conserving topsoil. His goal is to develop a healthy level of humus in the top two inches, which keeps the seed zone loose. He has stopped the sidewall compaction in the seed slot that still plagues his neighbours during wet springs. He attributes this improvement to the increase in humus and organic matter. His soil surface layer is crumbly and doesn't smear when the disk openers pass through. Bob proclaims that earthworms take the place of tillage by incorporating residue and converting it to humus. Worms help aerate his soil and improve internal drainage, which contributes to good rooting for his crops.

David Iles

On the Iles's North Carolina dairy farm the soil has actually changed from red to a dark, almost black colour since conversion to no-till in 1970.

David first learned about no-till from his college professor at North Carolina State University in 1964. Before he switched to no-till, David's corn silage yielded between 12 and 15 tons per acre in years with adequate rainfall and 4 to 5 tons in dry years—indicating that moisture was his major limiting factor. David realized that his water runoff losses and soil erosion were a direct result of tillage. Addressing the root cause of the problem, he switched to no-till and began to spread manure on 1/3 of his land annually. Since these changes, soil water is no longer limiting. With adequate rainfall he makes nearly 20 tons of silage now. David says his land is vastly more productive, with increased cation exchange capacity and increased phosphorus levels due to the humus present in his soil. Though his soil pH ranges in the 5.6 to 5.8 level, he applies no lime. His fields are more productive now than when he applied lime in the '70s and more productive than those of his neighbours who currently use lime and fertilizer.

David laments that this country has lost half of its topsoil in less than 100 years. orth Carolina State agronomist Bobby Brock agrees and says that for the first time in history we have the opportunity to produce food and build soil at the same time. David reasons that no-till is the way to improve the soil structure, increase tilth, and increase productivity while still practicing intensive agriculture. He realizes that organic matter is the engine that drives his system and provides food for earthworms and microorganisms. David built his soil by fallowing out 20 to 25 acres of his 380-acre farm each year. On these fallow acres he spreads manure and then sows crops that are not harvested but grown just for their organic matter. Even weeds are not clipped but left for their organic matter. David loves his earthworms and says they are the best employees he has. "They work all the time and eat dirt for a living."

His best field is one he cleared himself in the '70s. In spite cf traditional native pHs in the high 4s in his area, he did not lime this new ground but instead just planted rye on it. He had a fine rye crop that year, so he applied manure liberally and planted rye a second time. His second rye crop was excellent as well and was followed by corn the third year. That field yielded the most corn on the entire farm. This field has been in continuous corn since 1981 and has never been fertilized with conventional products or tilled. This field has a pH of 6.1 at a 6-inch depth, an exchange capacity of 8, and an 80% base saturation. David believes this field's productivity is high because it has never been harmed by tillage.

Earthworms and their Role in Vermicomposting

Vermicompost is nothing but the excreta of earthworms, which is rich in humus and nutrients. We can rear earthworms artificially in a brick tank or near the stem/trunk of trees (specially horticultural trees). By feeding these earthworms with biomass and watching properly the food (bio-mass) of earthworms, we can produce the required quantities of vermicompost.

Advantages of vermicompost:

1. Vermicompost is rich in all essential plant nutrients.
2. Provides excellent effect on overall plant growth, encourages the growth of new shoots/leaves and improves the quality and shelf life of the produce.
3. Vermicompost is free flowing, easy to apply, handle and store and does not have bad odour.
4. It improves soil structure, texture, aeration, and waterholding capacity and prevents soil erosion.
5. Vermicompost is rich in beneficial micro flora such as a fixers, P-solubilizers, cellulose decomposing micro-flora etc in addition to improve soil environment.
6. Vermicompost contains earthworm cocoons and increases the population and activity of earthworm in the soil.
7. It neutralizes the soil protection.
8. It prevents nutrient losses and increases the use efficiency of chemical fertilizers.
9. Vermicompost is free from pathogens, toxic elements, weed seeds etc.
10. Vermicompost minimizes the incidence of pest and diseases.
11. It enhances the decomposition of organic matter in soil.
12. It contains valuable vitamins, enzymes and hormones like auxins, gibberellins etc.

Composting as a recognized practice dates to at least the early Roman era since Pliny the Elder (AD 23-79) who refers to compost in his writings. Traditionally, composting was to pile organic materials until the next planting season, at which time the materials would be ready for soil application. The main advantage of this method is that little working time or effort is required from the composter and it fits in naturally with agricultural practices in temperate climates. Disadvantages (from the modern perspective) are that space is used for a whole year, some nutrients might be leached due to exposure to rainfall, and disease producing organisms, some weed, weed seeds and insects may not be adequately controlled.

Composting was somewhat modernized beginning in the 1960s in Europe as a tool for organic farming. The first industrial station for the transformation of urban organic materials into compost was set up in Wels/Austria in the year 1921. The early personages most cited for propounding composting within farming are for the German-speaking world Rudolf Steiner, founder of a farming method called biodynamics, and Annie France-Harrar, who was appointed on behalf of the government in Mexico and supported the country 1950–1958 to set up a large humus organization in the fight against erosion and soil degradation. In the English-speaking world it was Sir Albert Howard who worked extensively in India on sustainable practices and Lady Eve Balfour who was a huge proponent of composting. Composting was imported to America by various followers of these early European movements in the form of persons such as J.I. Rodale (founder of Rodale Organic Gardening), E.E. Pfeiffer (who developed scientific practices in biodynamic farming), Paul Keene (founder of Walnut Acres in Pennsylvania), and Scott and Helen Nearing (who inspired the back-to-land movement of the 1980s, Coincidentally, some of these personages met briefly in India- all were quite influential in the U.S. from the 1960s into the 1980s.

There are many modern proponents of rapid composting which attempt to correct some of the perceived problems associated with traditional, slow composting. Many advocate that compost can be made in 2 to 3 weeks. Many such short processes involve a few changes to traditional methods, including smaller, more homogenized pieces in the compost, controlling carbon to nitrogen (CN) ratio at 30 to 1 or less, and monitoring the moisture level more carefully.

However, none of these parameters differ significantly from early writings of Howard and Balfour, suggesting that in fact modern composting has not made significant advances over the traditional methods which take a few months to work. For this reason and others, many modern scientists who deal with carbon transformations are sceptical that there is a "super-charged" way to get nature to make compost rapidly. They also point to the fact that it is the structure of the natural molecules- such as carbohydrates, proteins, and cellulose- that really dictate the rate at which microbial-mediated transformations are possible.

Some cities such as San Francisco and Seattle require food and yard waste to be sorted for composting.

Ingredients

Composting organisms require four equally important things to work effectively:

- Carbon — for energy; the microbial oxidation of carbon produces the heat.
    - o High carbon materials tend to be brown and dry.

- Nitrogen — to grow and reproduce more organisms to oxidize the carbon.
    - o High nitrogen materials tend to be green (or colourful, such as fruits and vegetables) and wet.
- Oxygen — for oxidizing the carbon, the decomposition process.
- Water — in the right amounts to maintain activity without causing anaerobic conditions.

Certain ratios of these materials will provide beneficial bacteria with the nutrients to work at a rate that will heat up the pile. In that process much water will be released as [evaporation | vapour] ("steam"), and the oxygen will be quickly depleted, explaining the need to actively manage the pile. The hotter the pile gets, the more often added air and water is necessary; the air/water balance is critical to maintaining high temperatures until the materials are broken down. At the same time, too much air or water also slows the process, as does too much carbon (or too little nitrogen).

The most efficient composting occurs with a carbon:nitrogen mix of about 30 to 1. Nearly all plant and animal materials have both carbon and nitrogen, but amounts vary widely, with characteristics noted above (dry/wet, brown/green). Fresh grass clippings have an average ratio of about 15 to 1 and dry autumn leaves about 50 to 1 depending on species. Mixing equal parts by volume approximates the ideal C:N range. Few individual situations will provide the ideal mix of materials at any point in time- in this respect, home composting is like horseshoes, perfect is great, but close still works. Observation of amounts, and consideration of different materials as a pile is built over time, can quickly achieve a workable technique for the individual situation.

Urine

People excrete far more of certain water-soluble plant nutrients (nitrogen, phosphorus, potassium) in urine than in feces. Human urine can be used directly as fertilizer or it can be put onto compost. Adding a healthy person's urine to compost usually will increase temperatures and therefore increase its ability to destroy pathogens and unwanted seeds. Urine from a person with no obvious symptoms of infection, is generally much more sanitary than fresh feces. Unlike feces, urine doesn't attract disease-spreading flies (such as house flies or blow flies), and it doesn't harbor the most hardy of pathogens, such as parasitic worm eggs. Urine usually does not stink for long, particularly when it is fresh, diluted, or put on sorbents.

Urine is primarily composed of water and urea. Although metabolites of urea are nitrogen fertilizers, it is easy to over-fertilize with urine creating too much ammonia for plants to absorb, acidic conditions, or other phytotoxicity.

Micro-organisms

With the proper mixture of water, oxygen, carbon, and nitrogen, micro-organisms are allowed to break down organic matter to produce compost. The composting process is dependant on micro-organisms to break down organic matter into compost. There are many types of microorganisms found in active compost of which the most common are:

- Bacteria- The most numerous of all the microorganisms found in compost.
- Actinomycetes- Necessary for breaking down paper products such as newspaper, bark, etc.
- Fungi- Molds and yeast help break down materials that bacteria cannot, especially lignin in woody material.
- Protozoa- Help consume bacteria, fungi and micro organic particulates.
- Rotifers- Rotifers help control populations of bacteria and small protozoans.

In addition, earthworms not only ingest partly composted material, but also continually re-create aeration and drainage tunnels as they move through the compost.

A lack of a healthy micro-organisms community is the main reason why composting processes are slow in landfills with environmental factors such as lack of oxygen, nutrients or water being the cause of the depleted biological community.

Materials for Preparation of Vermicompost

Any types of biodegradable wastes-

1. Crop residues
2. Weed biomass
3. Vegetable waste
4. Leaf litter.
5. Hotel refuse
6. Waste from agro-industries
7. Biodegradable portion of urban and rural wastes.

Methods of Preparation

1. The vermicompost can be prepared in concrete tank. The size of the tank should be 10 ft. length or more depending upon the availability of land and raw materials, breadth 3- 5ft and height 3 ft. Suitable plastic tube/basin structure may also be needed. The

floor of the tank should be connected with stones and pieces of bricks.

2. The available bio-wastes are to be collected and are to be heaped under sun about 7-10 days and be chopped if necessary.
3. Sprinkling of cow dung slurry to the heap may be done.
4. A thin layer of half decomposed cow dung (1-2 inches) is to be placed at the bottom.
5. Place the chopped weed biomass and partially decomposed cow dung layer wise (10-20 cm) in the tank/pot upto the depth of 2 ½ ft. The bio waste and cow dung ratio should be 60: 40 on dry wt. Basis.
6. Release about 2-3 kg earthworms per ton of biomass or 100 nos. earthworms per one sq. ft. area.

Efficient Species

7. Place wire net/bamboo net over the tank to protect earthworm from birds.
8. Sprinkling of water should be done to maintain 70-80 % moisture content.
9. Provision of a shed over the compost is essential to prevent entry of rainwater and direct sunshine.
10. Sprinkling of water should be stopped when 90 % bio-wastes are decomposed. Maturity could be judged visually by observing the formation of granular structure of the compost at the surface of the tank.
11. Harvest the vermicompost by scrapping layer wise from the top of the tank and heap under shed. This will help in separation of earthworms from the compost. Sieving may also be done to separate the earthworms and cocoons.

Vermi Wash- a Plant growth Regulator

Vermi wash is a liquid plant growth regulator, which contains high amount of enzymes, vitamins and hormones like auxins, gibberellins etc along with macro and micronutrients used as foliar spray.

Methods of Preparation

1. Take one big bucket and one mug.
2. Set up one stop cork on the lower most part of the bucket.
3. Put a layer of broken bricks, pieces of stones having thickness of 10-15 cm in the bucket.

4. Over this layer put another layer of sand having thickness of 10-15 cm.
5. Then put a layer of partially decomposed cow dung having 30-45 cm thickness over it.
6. Then put another layer of soil having 2-3 thicknesses.
7. Now open the stopcock of the bucket and when the materials taken in the bucket.
8. Then put 100-200 nos. of earthworms in the bucket.
9. After that, a layer of paddy straw having 6 cm thickness is given.
10. Now open the stopcock of the bucket and spray water regularly for a period of 7-8 days.
11. After 10 days the liquid vermin wash will be produced in the bucket.
12. Hang one pot with a bottom hole over the bucket in such a way so that water falls drop by drop.
13. Every day 4-5 litres of water is to be poured in the hanging pot.
14. Keep another pot under stop cork to collect the vermin wash. Every day 3-4 litres vermin wash can be collected.

Application

1. Mix 1 litres of vermin wash with 7-10 litres of water and spray the solution in the leaf (upper and lower side) in the evening at the growing the crop.
2. Mix 1 litre of vermin wash with 1 litre of cow urine and then add 10 litres of water to the vermin urine solution and mixed thoroughly and keep it over night before spraying 50-60 litre of such solution and to be sprayed in one bigha of land to control various crop diseases.

Bio-fertilizers and their Utilization

Bio fertilizers are microbial inoculants consisting of living cells of micro-organism like bacteria, algae and fungi alone or combination which may help in increasing crop productivity by way of helping in the biological nitrogen fixation, solubilization of insoluble phosphorus fertilizer materials, stimulating growth or in decomposition of plant residues.

Advantages of using Bio-fertilizer

- Can replace 20 to 50 % of chemical fertilizer 'N' and 15 – 25 % of phosphatic fertilizers.
- Bio-fertilizers being cheap, provide highly cost effective supplement of chemical fertilizers.

- Increase farm productivity, generally 10-40 % in grain yield and 15 – 30 % in vegetable growth.
- Activates soil biologically thereby increasing natural fertility of soil, which causes sustainable agriculture.
- Help in stimulating plant growth in general and roots in particular as they serve various growth promoting hormones like auxin, gibberellins etc and vitamins. They help in better nutrient uptake and increase tolerance towards drought and moisture stress.
- Secrete some fungi static and antibiotic like substances, which reduce the incidence of various soil borne diseases. It has got antagonistic behaviour against pathogens.
- They help to proliferate and survival of other beneficial microorganisms in soil.
- Help in decomposing plant residues in soil, thereby improving C/N ratio, improving soil texture and structure. Also help in increasing water-holding capacity.
- In composting processes improve the quality and nutrient value of the compost.
- Being environment friendly and cost effective do not cause any harmful effect on soil and environment. On the contrary they help in degradation of other harmful chemicals such as weedicides and pesticides. Thereby they help in increasing the biodegradation capacity of the soil.

Types of Bio-fertilizers

For Nitrogen

a) Rhizobium bio-fertilizer to be used only in legume crops. Rhizobium in different for different legume crops.

b) Azatobacter/Azospirillum bio-fertilizer for all non-legume crops.

Azotobactum is for upland crop and Azospirillum is for low land crop.

For Phosphorus

Phosphatika for all the crops and to be used along with *Rhizobium/ Azotobacter a Azospirillum* bio- fertilizer.

Method of Application

There are three methods of application;

1. As seed treatment for direct seeded crop.
2. As seedling root dip Treatment for transplanted crops.
3. As soil treatment for potato, sugarcane crop where seed treatment and seedling.

7

Bio-remediation

Bioremediation is the use of microorganismal metabolism to remove pollutants. Technologies can be generally classified as *in situ* or *ex situ*. *In situ* bioremediation involves treating the contaminated material at the site, while *ex situ* involves the removal of the contaminated material to be treated elsewhere. Some examples of bioremediation technologies are bioventing, landfarming, bioreactor, composting, bioaugmentation, rhizofiltration, and biostimulation.

Bioremediation can occur on its own (natural attenuation or intrinsic bioremediation) or can be spurred on via the addition of fertilizers to increase the bioavailability within the medium (biostimulation). Recent advancements have also proven successful via the addition of matched microbe strains to the medium to enhance the resident microbe population's ability to break down contaminants. Microorganisms used to perform the function of bioremediation are known as Bioremediators.(bioaugmentation).

Not all contaminants, however, are easily treated by bioremediation using microorganisms. For example, heavy metals such as cadmium and lead are not readily absorbed or captured by organisms. The assimilation of metals such as mercury into the food chain may worsen matters. Phytoremediation is useful in these circumstances because natural plants or transgenic plants are able to bioaccumulate these toxins in their above-ground parts, which are then harvested for removal. The heavy metals in the harvested biomass may be further concentrated by incineration or even recycled for industrial use. The elimination of a wide range of pollutants and wastes from the environment requires increasing our understanding of the relative importance of different pathways and regulatory networks to carbon flux in particular environments and for particular compounds, and they will certainly accelerate the development of bioremediation technologies and biotransformation processes.

Genetic Engineering Approaches

The use of genetic engineering to create organisms specifically designed for bioremediation has great potential. The bacterium *Deinococcus radiodurans* (the most radioresistant organism known) has been modified to consume and digest toluene and ionic mercury from highly radioactive nuclear waste.

Most commonly, the process is misunderstood. The microbes are ever-present in any given context and are generally referred to as "normal microbial flora". During bioremediation (biodegradation) processes, fertilizer/nutrient supplementation is introduced to the environments in efforts to maximize growth and production potential. Common misbelief is that microbes are transported and dispersed into an unadulterated environment.

Mycoremediation

Mycoremediation is a form of bioremediation in which fungi are used to decontaminate the area. The term *mycoremediation* refers specifically to the use of fungal mycelia in bioremediation.

One of the primary roles of fungi in the ecosystem is decomposition, which is performed by the mycelium. The mycelium secretes extracellular enzymes and acids that break down lignin and cellulose, the two main building blocks of plant fibre. These are organic compounds composed of long chains of carbon and hydrogen, structurally similar to many organic pollutants. The key to mycoremediation is determining the right fungal species to target a specific pollutant. Certain strains have been reported to successfully degrade the nerve gases VX and sarin.

In one conducted experiment, a plot of soil contaminated with diesel oil was inoculated with mycelia of oyster mushrooms; traditional bioremediation techniques (bacteria) were used on control plots. After four weeks, more than 95% of many of the PAH (polycyclic aromatic hydrocarbons) had been reduced to non-toxic components in the mycelial-inoculated plots. It appears that the natural microbial community participates with the fungi to break down contaminants, eventually into carbon dioxide and water. Wood-degrading fungi are particularly effective in breaking down aromatic pollutants (toxic components of petroleum), as well as chlorinated compounds (certain persistent pesticides; Battelle, 2000).

Mycofiltration is a similar process, using fungal mycelia to filter toxic waste and microorganisms from water in soil.

Advantages

There are a number of cost/efficiency advantages to bioremediation, which can be employed in areas that are inaccessible without excavation. For example, hydrocarbon spills (specifically, petrol spills) or certain chlorinated solvents may contaminate groundwater, and introducing the appropriate electron acceptor or electron donor amendment, as appropriate, may significantly reduce contaminant concentrations after a long time allowing for acclimation. This is typically much less expensive than excavation followed by disposal elsewhere, incineration or other *ex situ* treatment strategies, and reduces or eliminates the need for "pump and treat", a practice common at sites where hydrocarbons have contaminated clean groundwater.

Monitoring Bioremediation

The process of bioremediation can be monitored indirectly by measuring the *Oxidation Reduction Potential* or redox in soil and groundwater, together with pH, temperature, oxygen content, electron acceptor/donor concentrations, and concentration of breakdown products (e.g. carbon dioxide). This table shows the (decreasing) biological breakdown rate as function of the redox potential.

Process	*Reaction*	*Redox potential (E_h in mV)*
aerobic:	$O_2 + 4e^- + 4H^+ \rightarrow 2H_2O$	600 ~ 400
anaerobic:		
denitrification	$2NO_3^- + 10e^- + 12H^+ \rightarrow N_2 + 6H_2O$	500 ~ 200
manganese IV reduction	$MnO_2 + 2e^- + 4H^+ \rightarrow Mn^{2+} + 2H_2O$	400 ~ 200
iron III reduction	$Fe(OH)_3 + e^- + 3H^+ \rightarrow Fe^{2+} + 3H_2O$	300 ~ 100
sulfate reduction	$SO_4^{2-} + 8e^- + 10\ H^+ \rightarrow H_2S + 4H_2O$	0 ~ –150
fermentation	$2CH_2O \rightarrow CO_2 + CH_4$	–150 ~ –220

This, by itself and at a single site, gives little information about the process of remediation.

1. It is necessary to sample enough points on and around the contaminated site to be able to determine contours of equal redox potential. Contouring is usually done using specialised software, e.g. using Kriging interpolation.
2. If all the measurements of redox potential show that electron acceptors have been used up, it is in effect an indicator for total microbial activity. Chemical analysis is also required to determine when the levels of contaminants and their breakdown products have been reduced to below regulatory limits.

Microbial Biodegradation

Interest in the microbial biodegradation of pollutants has intensified in recent years as humanity strives to find sustainable ways to cleanup contaminated environments. These bioremediation and biotransformation methods endeavour to harness the astonishing, naturally occurring, ability of microbial xenobiotic metabolism to degrade, transform or accumulate a huge range of compounds including hydrocarbons (e.g. oil), polychlorinated biphenyls (PCBs), polyaromatic hydrocarbons (PAHs), heterocyclic compounds (such as pyridine or quinoline), pharmaceutical substances, radionuclides and metals. Major methodological breakthroughs in recent years have enabled detailed genomic, metagenomic, proteomic, bioinformatic and other high-throughput analyses of environmentally relevant microorganisms providing unprecedented insights into key biodegradative pathways and the ability of organisms to adapt to changing environmental conditions.

The elimination of a wide range of pollutants and wastes from the environment is an absolute requirement to promote a sustainable development of our society with low environmental impact. Biological processes play a major role in the removal of contaminants and they take advantage of the astonishing catabolic versatility of microorganisms to degrade/convert such compounds. New methodological breakthroughs in sequencing, genomics, proteomics, bioinformatics and imaging are producing vast amounts of information. In the field of Environmental Microbiology, genome-based global studies open a new era providing unprecedented *in silico* views of metabolic and regulatory networks, as well as clues to the evolution of degradation pathways and to the molecular adaptation strategies to changing environmental conditions. Functional genomic and metagenomic approaches are increasing our understanding of the relative importance of different pathways and regulatory networks to carbon flux in particular environments and for particular compounds and they will certainly accelerate the development of bioremediation technologies and biotransformation processes.

Aerobic Biodegradation of Pollutants

The burgeoning amount of bacterial genomic data provides unparalleled opportunities for understanding the genetic and molecular bases of the degradation of organic pollutants. Aromatic compounds are among the most recalcitrant of these pollutants and lessons can be learned from the recent genomic studies of *Burkholderia xenovorans* LB400 and *Rhodococcus* sp. strain RHA1, two of the largest bacterial genomes completely sequenced to date. These studies have helped expand our understanding of bacterial catabolism, non-catabolic physiological adaptation to organic compounds, and the evolution of large bacterial genomes. First, the metabolic pathways from phylogenetically diverse isolates are very similar with respect to overall organization. Thus, as originally noted in pseudomonads, a large number of "peripheral aromatic" pathways funnel a range of natural and xenobiotic compounds into a restricted number of "central aromatic" pathways. Nevertheless, these pathways are genetically organized in genus-specific fashions, as exemplified by the b-ketoadipate and Paa pathways. Comparative genomic studies further reveal that some pathways are more widespread than initially thought. Thus, the Box and Paa pathways illustrate the prevalence of non-oxygenolytic ring-cleavage strategies in aerobic aromatic degradation processes. Functional genomic studies have been useful in establishing that even organisms harboring high numbers of homologous enzymes seem to contain few examples of true redundancy. For example, the multiplicity of ring-cleaving dioxygenases in certain rhodococcal isolates may be attributed to the cryptic aromatic catabolism of different terpenoids and steroids. Finally, analyses have indicated that recent genetic flux appears to have played a more significant role in the evolution of some large genomes, such as LB400's, than others. However, the emerging trend is that the large gene repertoires of potent pollutant degraders such as LB400 and RHA1 have evolved principally through more ancient processes. That this is true in such phylogenetically diverse species is remarkable and further suggests the ancient origin of this catabolic capacity.

Anaerobic Biodegradation of Pollutants

Anaerobic microbial mineralization of recalcitrant organic pollutants is of great environmental significance and involves intriguing novel biochemical reactions. In particular, hydrocarbons and halogenated compounds have long been doubted to be degradable in the absence of oxygen, but the isolation of hitherto unknown anaerobic hydrocarbon-degrading and reductively dehalogenating bacteria during the last decades provided ultimate proof for these processes in nature. Many novel biochemical reactions were discovered enabling the respective metabolic pathways, but progress in the molecular understanding of these bacteria was rather slow, since genetic systems are not readily applicable for most of them. However, with the increasing application of genomics in the field of environmental microbiology, a new and promising perspective is now at hand to obtain molecular insights into these new metabolic properties. Several complete genome sequences were determined during the last few years from bacteria capable of anaerobic organic pollutant degradation. 4.7 Mb genome of the facultative denitrifying *Aromatoleum aromaticum* strain EbN1 was the first to be determined for an anaerobic hydrocarbon degrader (using toluene or ethylbenzene as substrates).

The genome sequence revealed about two dozen gene clusters (including several paralogs) coding for a complex catabolic network for anaerobic and aerobic degradation of aromatic compounds. The genome sequence forms the basis for current detailed studies on regulation of pathways and enzyme structures. Further genomes of anaerobic hydrocarbon degrading bacteria were recently completed for the iron-reducing species *Geobacter metallireducens* (accession nr. NC_007517) and the perchlorate-reducing *Dechloromonas aromatica* (accession nr. NC_007298), but these are not yet evaluated in formal publications.

Complete genomes were also determined for bacteria capable of anaerobic degradation of halogenated hydrocarbons by halorespiration: the ~1.4 Mb genomes of *Dehalococcoides ethenogenes* strain 195 and *Dehalococcoides* sp. strain CBDB1 and the ~5.7 Mb genome of *Desulfitobacterium hafniense* strain Y51. Characteristic for all these bacteria is the presence of multiple paralogous genes for reductive dehalogenases, implicating a wider dehalogenating spectrum of the organisms than previously known. Moreover, genome sequences provided unprecedented insights into the evolution of reductive dehalogenation and differing strategies for niche adaptation.

Recently, it has become apparent that some organisms, including *Desulfitobacterium chlororespirans*, originally evaluated for halorespiration on chlorophenols, can also use certain brominated compounds, such as the herbicide bromoxynil and its major metabolite as electron acceptors for growth. Iodinated compounds may be dehalogenated as well, though the process may not satisfy the need for an electron acceptor.

Bioavailability, Chemotaxis, and Transport of Pollutants

Bioavailability, or the amount of a substance that is physiochemically accessible to microorganisms is a key factor in the efficient biodegradation of pollutants. O'Loughlin et al (2000) showed that, with the exception of kaolinite clay, most soil clays and cation exchange resins attenuated biodegradation of 2-picoline by *Arthrobacter* sp. strain R1, as a result of adsorption of the substrate to the clays. Chemotaxis, or the directed movement of motile organisms towards or away from chemicals in the environment is an important physiological response that may contribute to effective catabolism of molecules in the environment. In addition, mechanisms for the intracellular accumulation of aromatic molecules via various transport mechanisms are also important.

Oil Biodegradation

Petroleum oil contains aromatic compounds that are toxic for most life forms. Episodic and chronic pollution of the environment by oil causes major ecological perturbations. Marine environments are especially vulnerable since oil spills of coastal regions and the open sea are poorly containable and mitigation is difficult. In addition to pollution through human activities, about 250 million litres of petroleum enter the marine environment every year from natural seepages. Despite its toxicity, a considerable fraction of petroleum oil entering marine systems is eliminated by the hydrocarbon-degrading activities of microbial communities, in particular by a remarkable recently discovered group of specialists, the so-called hydrocarbonoclastic bacteria (HCB). *Alcanivorax borkumensis* was the first HCB to have its genome sequenced. In addition to hydrocarbons, crude oil often contains various heterocyclic compounds, such as pyridine, which appear to be degraded by similar, though separate mechanisms than hydrocarbons.

Analysis of Waste Biotreatment

Sustainable development requires the promotion of environmental management and a constant search for new technologies to treat vast quantities of wastes generated by increasing anthropogenic activities. Biotreatment, the processing of wastes using living organisms, is an environmentally friendly, relatively simple and cost-effective alternative to physico-chemical clean-up options. Confined environments, such as bioreactors, have been engineered to overcome the physical, chemical and biological limiting factors of biotreatment processes in highly controlled systems. The great versatility in the design of confined environments allows the treatment of a wide range of wastes under optimized conditions. To perform a correct assessment, it is necessary to consider various microorganisms having a variety of genomes and expressed transcripts and proteins. A great number of analyses are often required. Using traditional genomic techniques, such assessments are limited and time-consuming. However, several high-throughput techniques originally developed for medical studies can be applied to assess biotreatment in confined environments.

Metabolic Engineering and Biocatalytic Applications

The study of the fate of persistent organic chemicals in the environment has revealed a large reservoir of enzymatic reactions with a large potential in preparative organic synthesis, which has already been exploited for a number of oxygenases on pilot and even on industrial scale. Novel catalysts can be obtained from metagenomic libraries and DNA sequence based approaches. Our increasing capabilities in adapting the catalysts to specific reactions and process requirements by rational and random mutagenesis broadens the scope for application in the fine chemical industry, but also in the field of biodegradation. In many cases, these catalysts need to be exploited in whole cell bioconversions or in fermentations, calling for system-wide approaches to understanding strain physiology and metabolism and rational approaches to the engineering of whole cells as they are increasingly put forward in the area of systems biotechnology and synthetic biology.

Fungal Biodegradation

In the ecosystem, different substrates are attacked at different rates by consortia of organisms from different kingdoms. *Aspergillus* and other moulds play an important role in these consortia because they are adept at recycling starches, hemicelluloses, celluloses, pectins and other sugar polymers. Some aspergilli are capable of degrading more refractory compounds such as fats, oils, chitin, and keratin. Maximum decomposition occurs when there is sufficient nitrogen, phosphorus and other essential inorganic nutrients. Fungi also provide food for many soil organisms. For *Aspergillus* the process of degradation is the means of obtaining nutrients. When these moulds degrade human-made substrates, the process usually is called biodeterioration. Both paper and textiles (cotton, jute, and linen) are particularly vulnerable to *Aspergillus* degradation. Our artistic heritage is also subject to *Aspergillus* assault. To give but one example, after Florence in Italy flooded in 1969, 74% of the isolates from a damaged Ghirlandaio fresco in the Ognissanti church were *Aspergillus versicolor.*

Phyto-remediation

Phytoremediation, and Latin *remedium* (*restoring balance* or *remediation*) describes the treatment of environmental problems (bioremediation) through the use of plants that mitigate the environmental problem without the need to excavate the contaminant material and dispose of it elsewhere.

Phytoremediation consists in mitigating pollutant concentrations in contaminated soils, water, or air, with plants able to contain, degrade, or eliminate metals, pesticides, solvents, explosives, crude oil and its derivatives, and various other contaminants from the media that contain them.

Application

Phytoremediation may be applied wherever the soil or static water environment has become polluted or is suffering ongoing chronic pollution. Examples where phytoremediation has been used successfully include the restoration of abandoned metal-mine workings, reducing the impact of sites where polychlorinated biphenyl have been dumped during manufacture and mitigation of on-going coal mine discharges.

Phytoremediation refers to the natural ability of certain plants called hyperaccumulators to bioaccumulate, degrade,or render harmless contaminants in soils, water,or air. Contaminants such as metals, pesticides, solvents, explosives, and crude oil and its derivatives, have been mitigated in phytoremediation projects worldwide. Many plants such as mustard plants, alpine pennycress, and pigweed have proven to be successful at hyperaccumulating contaminants at toxic-waste sites.

Phytoremediation is considered a clean, cost-effective and non-environmentally disruptive technology, as opposed to mechanical cleanup methods such as soil excavation or pumping polluted groundwater. Over the past 20 years, this technology has become increasingly popular and has been employed at sites with soils contaminated with lead, uranium, and arsenic. However, one major disadvantage of phytoremediation is that it requires a long-term commitment, as the process is dependent on plant growth, tolerance to toxicity, and bioaccumulation capacity.

Advantages and Limitations

- Advantages:
  - o the cost of the phytoremediation is lower than that of traditional processes both *in situ* and *ex situ*
  - o the plants can be easily monitored
  - o the possibility of the recovery and re-use of valuable metals (by companies specializing in "phyto mining")
  - o it is potentially the least harmful method because it uses naturally occurring organisms and preserves the environment in a more natural state.
- Limitations:
  - o phytoremediation is limited to the surface area and depth occupied by the roots.
  - o slow growth and low biomass require a long-term commitment
  - o with plant-based systems of remediation, it is not possible to completely prevent the leaching of contaminants into the groundwater (without the complete removal of the contaminated ground, which in itself does not resolve the problem of contamination)
  - o the survival of the plants is affected by the toxicity of the contaminated land and the general condition of the soil.
  - o bio-accumulation of contaminants, especially metals, into the plants which then pass into the food chain, from primary level consumers upwards and/or requires the safe disposal of the affected plant material.

Various Phytoremediation Processes

A range of processes mediated by plants or algae are useful in treating environmental problems:

- *Phytoextraction* — uptake and concentration of substances from the environment into the plant biomass.
- *Phytostabilization* — reducing the mobility of substances in the

environment, for example, by limiting the leaching of substances from the soil.

- *Phytotransformation* — chemical modification of environmental substances as a direct result of plant metabolism, often resulting in their inactivation, degradation (phytodegradation), or immobilization (phytostabilization).
- *Phytostimulation* — enhancement of soil microbial activity for the degradation of contaminants, typically by organisms that associate with roots. This process is also known as *rhizosphere degradation*. Phytostimulation can also involve aquatic plants supporting active populations of microbial degraders, as in the stimulation of atrazine degradation by hornwort.
- *Phytovolatilization* — removal of substances from soil or water with release into the air, sometimes as a result of phytotransformation to more volatile and/or less polluting substances.
- *Rhizofiltration* — filtering water through a mass of roots to remove toxic substances or excess nutrients. The pollutants remain absorbed in or adsorbed to the roots.

Phytoextraction

Phytoextraction (or *phytoaccumulation*) uses plants or algae to remove contaminants from soils, sediments or water into harvestable plant biomass. Phytoextraction has been growing rapidly in popularity worldwide for the last twenty years or so. In general, this process has been tried more often for extracting heavy metals than for organics. At the time of disposal, contaminants are typically concentrated in the much smaller volume of the plant matter than in the initially contaminated soil or sediment. 'Mining with plants', or phytomining, is also being experimented with.

The plants absorb contaminants through the root system and store them in the root biomass and/or transport them up into the stems and/or leaves. A living plant may continue to absorb contaminants until it is harvested. After harvest, a lower level of the contaminant will remain in the soil, so the growth/harvest cycle must usually be repeated through several crops to achieve a significant cleanup. After the process, the cleaned soil can support other vegetation.

Advantages: The main advantage of phytoextraction is environmental friendliness. Traditional methods that are used for cleaning up heavy metal-contaminated soil disrupt soil structure and reduce soil productivity, whereas phytoextraction can clean up the soil without causing any kind of harm to soil quality. Another benefit of phytoextraction is that it is less expensive than any other clean-up process.

Disadvantages: As this process is controlled by plants, it takes more time than anthropogenic soil clean-up methods.

Two versions of phytoextraction:

- natural hyper-accumulation, where plants naturally take up the contaminants in soil unassisted, and
- induced or assisted hyper-accumulation, in which a conditioning fluid containing a chelator or another agent is added to soil to increase metal solubility or mobilization so that the plants can absorb them more easily. In many cases natural hyperaccumulators are metallophyte plants that can tolerate and incorporate high levels of toxic metals.

Examples of phytoextraction:

- Arsenic, using the Sunflower (*Helianthus annuus*), or the Chinese Brake fern (*Pteris vittata*), a hyperaccumulator. Chinese Brake fern stores arsenic in its leaves.
- Cadmium, using Willow (*Salix viminalis*): In 1999, one research experiment performed by Maria Greger and Tommy Landberg suggested Willow (Salix viminlais) has a significant potential as a phytoextractor of Cadmium (Cd), Zinc (Zn), and Copper (Cu), as willow has some specific characteristics like high transport capacity of heavy metals from root to shoot and huge amount of biomass production; can be used also for production of bio energy in the biomass energy power plant.
- Cadmium and zinc, using Alpine pennycress (*Thlaspi caerulescens*), a hyperaccumulator of these metals at levels that would be toxic to many plants. On the other hand, the presence of copper seems to impair its growth.
- Lead, using Indian Mustard (*Brassica juncea*), Ragweed (*Ambrosia artemisiifolia*), Hemp Dogbane (*Apocynum cannabinum*), or Poplar trees, which sequester lead in its biomass.
- Salt-tolerant (moderately halophytic) barley and/or sugar beets are commonly used for the extraction of Sodium chloride (common salt) to reclaim fields that were previously flooded by sea water.
- Caesium-137 and strontium-90 were removed from a pond using sunflowers after the Chernobyl accident.
- Mercury, selenium and organic pollutants such as polychlorinated biphenyls (PCBs) have been removed from soils by transgenic plants containing genes for bacterial enzymes.

Phytostabilization

Phytostabilization focuses on long-term stabilization and containment of the pollutant. For example, the plant's presence can reduce wind erosion; or the plant's roots can prevent water erosion, immobilize the pollutants by adsorption or accumulation, and provide a zone around the roots where the pollutant can precipitate and stabilize. Unlike phytoextraction, phytostabilization focuses mainly on sequestering pollutants in soil near the roots but not in plant tissues. Pollutants become less bioavailable, and livestock, wildlife, and human exposure is reduced. An example application of this sort is using a vegetative cap to stabilize and contain mine tailings.

Phytotransformation

In the case of organic pollutants, such as pesticides, explosives, solvents, industrial chemicals, and other xenobiotic substances, certain plants, such as Cannas, render these substances non-toxic by their metabolism. In other cases, microorganisms living in association with plant roots may metabolize these substances in soil or water. These complex and recalcitrant compounds cannot be broken down to basic molecules (water, carbon-dioxide, etc.) by plant molecules, and, hence, the term *phytotransformation* represents a change in chemical structure without complete breakdown of the compound. The term "Green Liver Model" is used to describe phytotransformation, as plants behave analogously to the human liver when dealing with these xenobiotic compounds (foreign compound/ pollutant). After uptake of the xenobiotics, plant enzymes increase the polarity of the xenobiotics by adding functional groups such as hydroxyl groups (-OH).

This is known as Phase I metabolism, similar to the way that the human liver increases the polarity of drugs and foreign compounds (Drug Metabolism). Whereas in the human liver enzymes such as Cytochrome P450s are responsible for the initial reactions, in plants enzymes such as nitroreductases carry out the same role.

In the second stage of phytotransformation, known as Phase II metabolism, plant biomolecules such as glucose and amino acids are added to the polarized xenobiotic to further increase the polarity (known as conjugation). This is again similar to the processes occurring in the human pancreas where glucuronidation (addition of glucose molecules by the UGT (e.g. UGT1A1) class of enzymes) and glutathione addition reactions occur on reactive centres of the xenobiotic.

Phase I and II reactions serve to increase the polarity and reduce the toxicity of the compounds, although many exceptions to the rule are seen. The increased polarity also allows for easy transport of the xenobiotic along aqueous channels.

In the final stage of phytotransformation (Phase III metabolism), a sequestration of the xenobiotic occurs within the plant. The xenobiotics polymerize in a lignin-like manner and develop a complex structure that is sequestered in the plant. This ensures that the xenobiotic is safely stored, and does not affect the functioning of the plant. However, preliminary studies have shown that these plants can be toxic to small animals (such as snails), and, hence, plants involved in phytotransformation may need to be maintained in a closed enclosure. Hence, the plants reduce toxicity (with exceptions) and sequester the xenobiotics in phytotransformation. Trinitrotoluene phytotransformation has been extensively researched and a transformation pathway has been proposed.

The Role of Genetics

Breeding programs and genetic engineering are powerful methods for enhancing natural phytoremediation capabilities, or for introducing new capabilities into plants. Genes for phytoremediation may originate from a micro-organism or may be transferred from one plant to another variety better adapted to the environmental conditions at the cleanup site. For example, genes encoding a nitroreductase from a bacterium were inserted into tobacco and showed faster removal of TNT and enhanced resistance to the toxic effects of TNT. Researchers have also discovered a mechanism in plants that allows them to grow even when the pollution concentration in the soil is lethal for non-treated plants. Some natural, biodegradable compounds, such as exogenous polyamines, allow the plants to tolerate concentrations of pollutants 500 times higher than untreated plants, and to absorb more pollutants.

Hyperaccumulators and Biotic Interactions

A plant is said to be a hyperaccumulator if it can concentrate the pollutants in a minimum percentage which varies according to the pollutant involved (for example: more than 1000 mg/kg of dry weight for nickel, copper, cobalt, chromium or lead; or more than 10,000 mg/kg for zinc or manganese). This capacity for accumulation is due to hypertolerance, or *phytotolerance*: the result of adaptative evolution from the plants to hostile environments through many generations. A number of interactions may be affected by metal hyperaccumulation, including protection, interferences with neighbour plants of different species, mutualism (including mycorrhizae, pollen and seed dispersal), commensalism, and biofilm.

Phytotreatment

Phytotreatment is an environmental term referring to the cultivation of specialized plants that absorb specific contaminants from the soil through their roots or foliage. This reduces the concentration of contaminants in the soil, but incorporates them into biomasses that may be released back into the environment when the plant dies or is harvested.

8

Bioleaching by Microbial Process

Bioleaching is the extraction of specific metals from their ores through the use of bacteria. This is much cleaner than the traditional heap leaching using cyanide. Bioleaching is one of several applications within biohydrometallurgy and several methods are used to recover copper, zinc, lead, arsenic, antimony, nickel, molybdenum, gold, and cobalt.

The Process

The extraction of iron can involve numerous ferrous and sulfur oxidizing bacteria, including *Acidithiobacillus ferrooxidans* and *Acidithiobacillus thiooxidans* (formerly known as *Thiobacillus*). For example, bacteria catalyse the breakdown of the mineral arsenopyrite (FeAsS) by oxidising the sulfur and metal (in this case arsenic ions) to higher oxidation states whilst reducing dioxygen by H_2 and Fe^{3+}. This allows the soluble products to dissolve.

$$FeAsS_{(s)} \rightarrow Fe^{2+}_{(aq)} + As^{3+}_{(aq)} + S^{6+}_{(aq)}$$

This process actually occurs at the period of when the [cell membrane]] of the bacteria. The electrons pass into the cells and are used in biochemical processes to produce energy for the bacteria to reduce oxygen molecules to water.

In stage 2, bacteria oxidise Fe^{2+} to Fe^{3+} (whilst reducing O_2).

$$Fe^{2+} \rightarrow Fe^{3+}$$

They then oxidise the metal to a higher positive oxidation state. With the electrons gained, they reduce Fe^{3+} to Fe^{2+} to continue the cycle.

$$M^{3+} \rightarrow M^{5+}$$

The iron is now separated from the ore and in solution.

The process for copper is very similar however the efficiency and kinetics depend on the copper mineralogy. The most efficient minerals are supergene minerals such as chalcocite, Cu_2S and Covellite, CuS. The main copper mineral chalcopyrite ($CuFeS_2$) is not leached very efficiently which is why the dominant copper producing technology remains flotation followed by smelting and refining. The leaching of CuFeS2 follows the two stages of being dissolved and then further oxidised, with Cu^{2+} ions being left.

Extraction from Mixture

Copper (Cu^{2+}) ions are removed from the solution by ligand exchange solvent extraction which leaves other ions in the solution. The copper is removed by bonding to a ligand, which is a large molecule consisting of a number of smaller groups each possessing a lone pair. The ligand is dissolved in an organic solvent such as kerosene and shaken with the solution producing this reaction:

$$Cu^{2+}_{(aq)} + 2LH(organic) \rightarrow CuL_2(organic) + 2H^+_{(aq)}$$

The ligand donates electrons to the copper, producing a complex- a central metal atom (copper) bonded to 2 molecules of the ligand. Because this complex has no charge, it is no longer attracted to polar water molecules and dissolves in the kerosene, which is then easily separated from the solution. Because the initial reaction is reversible, it is determined by pH. Adding concentrated acid reverses the equation, and the copper ions go back into an aqueous solution.

Then the copper is passed through an electro-winning process to increase its purity: an electric current is passed through the resulting solution of copper ions. Because copper ions have a 2+ charge, they are attracted to the negative cathodes and collect there.

The copper can also be concentrated and separated by displacing the copper with Fe from scrap iron:

$$Cu^{2+}_{(aq)} + Fe_{(s)} \rightarrow Cu_{(s)} + Fe^{2+}_{(aq)}$$

The electrons lost by the iron are taken up by the copper. Copper is the oxidising agent (it accepts electrons), and iron is the reducing agent (it loses electrons).

Traces of precious metals such as gold may be left in the original solution. Treating the mixture with sodium cyanide in the presence of free oxygen dissolves the gold. The gold is removed from the solution by adsorbing (taking it up on the surface) to charcoal.

Bioleaching with Fungi

Several species of fungi can be used for bioleaching. Fungi can be grown on many different strata, as with electronic scrap, catalytic converters, and fly ash from municipal waste incineration. Experiments have shown that two fungal strains (*Aspergillus Niger, Penicillium simplicissimum*) were able to mobilize Cu and Sn by 65%, and Al, Ni, Pb, and Zn by more than 95%.*Aspergillus Niger* can produce some organic acids such as citric acid. So it can be used for bioleaching sulfides.

Compared with other Extraction Techniques

Traditional extractions involve many expensive steps such as roasting and smelting, which require sufficient concentrations of elements in ores and are environmentally unfriendly. Low concentrations are not a problem for bacteria because they simply ignore the waste which surrounds the metals, attaining extraction yields of over 90% in some cases. These microorganisms actually gain energy by breaking down minerals into their constituent elements. The company simply collects the ions out of the solution after the bacteria have finished.

Some advantages associated with bioleaching are:

- economical: bioleaching is generally simpler and therefore cheaper to operate and maintain than traditional processes, since fewer specialists are needed to operate complex chemical plants.
- environmental: The process is more environmentally friendly than traditional extraction methods. For the company this can translate into profit, since the necessary limiting of sulfur dioxide emissions during smelting is expensive. Less landscape damage occurs, since the bacteria involved grow naturally, and the mine and surrounding area can be left relatively untouched. As the bacteria breed in the conditions of the mine, they are easily cultivated and recycled.

Some disadvantages associated with bioleaching are:

- economical: the bacterial leaching process is very slow compared to smelting. This brings in less profit as well as introducing a significant delay in cash flow for new plants.
- environmental: Toxic chemicals are sometimes produced in the process. Sulfuric acid and H^+ ions which have been formed can leak into the ground and surface water turning it acidic, causing environmental damage. Heavy ions such as iron, zinc, and arsenic leak during acid mine drainage. When the pH of this solution rises, as a result of dilution by fresh water, these ions precipitate, forming "Yellow Boy" pollution. For these reasons, a setup of bioleaching must be carefully planned, since the process can lead to a biosafety failure.

Currently it is more economical to smelt copper ore rather than to use bioleaching, since the concentration of copper in its ore is generally quite high. The profit obtained from the speed and yield of smelting justifies its cost. However, the concentration of gold in its ore is generally very low.

The lower cost of bacterial leaching in this case outweighs the time it takes to extract the metal.

Biocatalytic Processes

The physiology and energetic manipulation of metal leaching organisms are extremely significant, especially for gold, copper, uranium leaching and recovery of metals such as arsenic, silver, and mercury. Organisms that thrive in extreme environments such as those described above, are of interest in the production of highly stable enzymes and in the development of certain innovative bioprocesses. One area of environmental/ biotechnological research would be the realization of the biocatalytic potential of these extremophilic microbes, which thrive at very high (boiling) temperatures, high pressures, highly saline or acidic environments. An area of interest would be the development of environmentally relevant (bio)technology based on microbial degradation of the recalcitrant pollutants. This calls for characterization of single microbes and mixed cultures that can survive amidst high concentrations of the pollutants. Essential evaluation of microbial physiology in conjunction with industrially relevant molecular bioprocess design is required for development of these systems which can aid as great learning/research tools in arenas of enzyme or cell processing applications along with mining, waste-water treatment, bioremediation. Understanding of these "hyperthermophilic anaerobes" that encompass a wide metabolic variety can be utilized for novel applications such as high-temperature 'anaerobic digestors', aiding the conversion of the waste to useful products, molecular engineering of enzymes etc.

Bacterial Oxidation

Bacterial oxidation (BIOX) is a biohydrometallurgical process developed for pre-cyanidation treatment of refractory gold ores or concentrates. The bacterial culture is a mixed culture of *Thiobacillus ferrooxidans*, *Thiobacillus thiooxidans* and *Leptospirillum ferrooxidans*. The bacterial oxidation process comprises contacting refractory sulfide ROM ore or concentrate with a strain of the bacterial culture for a suitable treatment period under an optimum operating environment. The bacteria oxidise the sulfide minerals, thus liberating the occluded gold for subsequent recovery via cyanidation.

Under controlled continuous plant conditions, the number of bacterial cells and their activity is optimised to attain the highest rate of sulfide oxidation. The bacteria require a very acidic environment (pH 1.0 to 4.0), a temperature of between 30 and 45 °C, and a steady supply of oxygen and carbon dioxide for optimum growth and activity. The unusual operating conditions for the bacteria are not favourable for the growth of most other microbes, thus eliminating the need for sterility during the bacterial oxidation process. Because organic substances are toxic to the bacteria, they are non-pathogenic and incapable of causing disease. The bacteria employed in the process do not, therefore, pose a health risk to humans or any animals.

The bacterial oxidation of iron sulfide minerals produces iron(III) sulfate and sulfuric acid, and in the case of arsenopyrite, arsenic acid is also produced. The arsenic is removed from the liquor by coprecipitation with the iron and sulfate in a two stage neutralisation process. This produces a solid neutralisation precipitate containing largely calcium sulfate, basic iron(III) arsenate and iron(III) hydroxide. The iron(III) arsenate is sufficiently insoluble and stable to allow the neutralisation product to be safely disposed of on a slimes dam. The neutralisation liquor, purified to contain an acceptable level of arsenic, can be re-used in the milling, flotation or bacterial oxidation circuits.

Integrated Nutrient Management Approach in Wetland Rice Ecosystem

More than 40 percent of the world's population depends on rice for its major source of food. In order to feed a growing population, annual world rice production needs to increase from 586 million tonnes (Mt) (2001) to 756 Mt by 2030 (FAO, 2000a). The need for such an increase is reflected in efforts to: increase the acreage under high-yielding varieties (HYV); develop hybrids and super rice; evolve more appropriate and efficient crop, soil, water and nutrient management technologies; and accelerate technology transfer. Rice is grown in various climates ranging from tropical and subtropical to temperate, with the tropics the major area of cultivation. Within their seven broadly-defined agro-ecozones, rice systems are further characterized and quantified in terms of their rice-phase water regimes: irrigated (56%); rainfed lowland (26%); flood-prone/swampland (9%); and rainfed upland (9%). The irrigated sector generates more than 75 percent of total rice production. The estimated global rice acreage is about 152 million ha (Mha)- an increase of 3.5 percent since 1990.

Table : Rice area, production and yield

	1990	*1995*	*2001*
Asia			
Area (*Mha*)	132.3	133.6	136.4
Production (*Mt*)	477.6	499.2	532.8
Yield (*kg/ha*)	3 610	3 737	3 906
World			
Area (*Mha*)	146.85	149.44	152.04
Production (*Mt*)	518.16	547.10	585.59
Yield (*kg/ha*)	3 528	3 661	3 851

Rice is Asia's main crop, and the region accounts for 90 percent of the world's acreage and produces 91 percent of the world's rice production. About 250 million households in Asia (70% of the world's farming households) cultivate rice and support 2 600 million rice consumers. Within Asia, rice occupies about 25 percent of all land under arable and permanent crops. Per caput rice area in Asia fell from 0.12 ha in 1961 to 0.07 ha in 1998. The average holding size also declined during this period, partly due to land reform/redistribution programmes, and partly due to inheritance-related over-fragmentation of smallholdings. National average yields range from 2.6 t/ha in Thailand to 6.4 t/ha in China (2001).

Rice Production Projections to 2015 and 2030

FAO estimates of crop yields and area for 2015 and 2030 rely on country forecasts of demand and supply variables. The most important demand determinants are population, income, commodity prices and dietary tastes and preferences. Food supply is assumed to be mainly a function of cropped area and yield, but crop and input prices and other agricultural policies can play a determining role in planted acreage and output yield.

In Asia, the potential for expanding crop area is limited in most countries. Moreover, there has been a decline in the rate of increase in irrigated area. The overall area is expected to increase from 135 Mha in 1995-97 (average taken as base year) to 147 Mha in 2030. Although a reduction in rice acreage is projected for Japan, the Republic of Korea, Malaysia, Thailand and the Philippines, overall rice production needs to increase from 516 Mt in 1995-97 to 645 Mt in 2015 and 694 Mt in 2030. Production is expected to rise in all countries (with the exception of Japan, the Republic of Korea and Malaysia). The projected slowdown in overall production growth for 2015-2030 reflects the anticipated deceleration in population growth and the change towards a dietary preference for meat.

Table : Rice area, yield and production projections for 2015 and 2030

Country	*Area (Mha)*			*Yield (t/ha)*			*Production (Mt)*		
	1995-97	*2015*	*2030*	*1995-97*	*2015*	*2030*	*1995-97*	*2015*	*2030*
India	43.1	48.0	49.2	2.8	3.9	4.2	122.2	187.5	205.6
China	31.3	34.8	34.0	6.2	6.1	6.3	193.6	211.5	212.6
Indonesia	11.4	12.2	12.5	4.4	5.1	5.4	50.0	61.9	67.3
Bangladesh	10.1	11.0	11.4	2.7	3.3	3.7	27.6	36.7	42.0
Thailand	9.4	8.9	8.3	2.4	2.8	3.2	22.6	24.7	26.5
Viet Nam	7.0	7.6	7.8	3.8	5.0	5.5	26.3	37.7	43.1
Myanmar	5.8	6.5	7.2	3.1	3.3	3.6	17.8	21.8	25.9
Philippines	3.9	3.8	3.8	2.9	3.8	4.2	11.3	14.7	16.0
Pakistan	2.2	2.5	2.8	2.8	3.7	4.5	6.3	9.4	12.4

Japan	2.0	1.5	1.2	6.4	7.2	7.7	13.0	10.6	9.2
Cambodia	1.9	2.4	2.7	1.8	2.3	2.7	3.4	5.5	7.5
Nepal	1.5	1.6	1.6	2.4	3.0	3.4	3.6	4.8	5.7
Rep. of Korea	1.1	0.9	0.8	6.6	6.9	7.2	6.9	6.1	5.4
Sri Lanka	0.7	0.7	0.7	3.2	4.4	4.8	2.4	3.2	3.4
Malaysia	0.7	0.6	0.6	3.2	3.5	3.6	2.2	2.2	2.2
Lao PDR	0.6	0.7	0.8	2.6	3.3	3.9	1.5	2.4	3.2
Iran	0.6	0.9	1.1	4.1	4.7	5.1	2.4	4.1	5.6
Asia	134.9	144.6	146.6	3.8	4.5	4.7	516.3	644.8	693.6

As in Japan, the Republic of Korea, Malaysia, Thailand and the Philippines, other countries are likely to transfer more land currently under rice (especially marginal land and areas vulnerable to degradation) to other uses. Yield increases in favourable areas will be the main element for output growth. Hence, current yield potentials and yield gaps warrant careful analysis. Gaps between potential and current yield levels are substantial. The irrigated ecology offers the greatest scope for narrowing the substantial gaps between potential and current yield levels; indeed, the projected demand and production potential for 2030 may be within reach.

Lin and Shen (1996) analysed yield gaps under favourable conditions in China. Technical constraints accounted for between 41 and 45 percent of the yield differences, with deficiencies in N, P, K, soil organic matter and micronutrients the most important. This situation also reflected the long history of intensive cropping, soil fertility depletion and an inability to recover through natural processes. In India, the major issues and challenges for bridging yield gaps (Siddiq, 2000) pertain to:

- yield plateauing in high productivity areas;
- declining productivity due to inappropriate soil management practices in major rice-based cropping systems under irrigated ecologies;
- continued imbalanced fertilizer use;
- widespread deficiencies in micro- and secondary nutrients, such as zinc and sulphur;
- low input management vis-à-vis risk of crop losses; and
- lack of varietal solutions to problematic soils.

Thus, improved nutrient management is essential for the further exploitation of the genetic yield potentials of HYVs.

Nutrient Requirements for Projected Production Targets

FAO uses relatively straightforward techniques to obtain mineral fertilizer use estimates (FAO, 2000b). The increased nutrient-use efficiency scenario is probably more appropriate for rice because of the immense scope for improving N-use efficiency. Ideally, there should be a crop-based model for each country, as climate, resource, production and other factors vary considerably. However, such an endeavour would be fraught with data limitations. In the absence of known crop production functions, estimates of future fertilizer application rates and nutrient-use efficiencies were obtained by quantifying the relationship between production and total fertilizer application rates that existed in the period 1995-97 for all producing countries. This global crop-specific fertilizer response coefficient was used to compute the future fertilizer application rates required to attain the projected yield increases.

The projections reveal that N use per hectare would increase in all Asian countries over the years. Total N consumption is also projected to increase in all countries, except in China, Japan, the Republic of Korea and Malaysia, where a significant decrease in total consumption is expected, mainly due to the reduction in area. The aggregated projected nitrogen use figures for Asia for 2015 and 2030 are 16.0 and 16.6 Mt, respectively.

Table : Projections for nitrogen fertilizer use

Country	*N use (kg/ha)*			*Total N use ('000 tonnes)*		
	1995-97	*2015*	*2030*	*1995-97*	*2015*	*2030*
India	94	129	136	4 037	5 013	5 301
China	144	179	184	5 702	5 410	5 349
Indonesia	145	167	176	1 649	1 836	1 919
Bangladesh	80	97	105	802	939	1 021
Thailand	56	65	74	528	533	536
Viet Nam	118	155	169	821	986	1 068
Myanmar	21	23	24	121	146	171
Philippines	58	76	83	223	259	271
Pakistan	108	142	166	242	289	341
Japan	88	98	106	178	140	118
Cambodia	4	5	6	7	12	16
Nepal	17	21	24	25	32	37
Rep. of Korea	178	187	194	187	158	138
Sri Lanka	95	133	142	71	78	79
Malaysia	92	101	103	63	58	58
Lao PDR	3	4	5	2	3	4
Iran	133	150	163	78	120	153
Asia	14	735	16	012	16	581

Nitrogen-use Efficiency

The calculation of N-use efficiency assumes that 20 kg of N is needed for the production of 1 tonne of rice and that applied fertilizer provides 50 percent of this requirement.

Average N-use efficiency in Asia is projected to rise from 35 percent in 1995-97 to 40.3 percent by 2015 and 41.8 percent by 2030. Although N-use efficiency is less than 35 percent in almost all Asian countries, except for Japan, Philippines, Thailand and Republic of Korea, there is ample scope for improvement through the pooling of research and development and technology transfer resources.

Table : Estimates of N-use efficiency in selected countries

Country	*Nitrogen-use efficiency (%)*		
	1995-97	*2015*	*2030*
India	30.3	37.4	38.8
China	34.0	39.1	39.7
Indonesia	30.3	33.7	35.1
Bangladesh	34.4	39.1	41.1
Thailand	42.8	46.3	49.4
Viet Nam	32.0	38.2	40.4
Philippines	50.7	56.8	59.0
Pakistan	26.0	32.5	36.4
Japan	73.0	75.7	78.0
Rep. of Korea	36.9	38.6	39.1
Sri Lanka	33.8	41.0	43.0
Malaysia	34.9	37.9	37.9
Iran	30.8	34.2	36.6
Asia	35.0	40.3	41.8

A major cause of low N-use efficiency is its loss from the plant-soil system through gaseous emissions, runoff, erosion and leaching. The relative importance of each of these pathways varies from site to site. The adjustment of N fertilizer inputs in accordance with plant demand and appropriate application methods reduces such losses. Gaseous N loss processes include volatilization and denitrification, resulting in the release of NH_3, and NO, N_2O and N_2 to the atmosphere. Factors such as climate and soil-crop-water management contribute to regional variations. For Asia, the estimated total NH3 loss from mineral N application in wetland rice for 1995 was 2.2 Mt.

Table : Estimated NH3 volatilization loss from mineral N applications in wetland rice in Asia, 1995

Region (Mha)	*Area ('000 tonnes)*	*Mineral N use ('000 tonnes)*	*NH_3-N loss (%)*	*Percent loss*
South Asia	44	4 646	1 030	22.2
East Asia	26	4 490	829	18.5
Southeast Asia	35	1 811	335	18.5
Japan	2	171	24	14.0

Source: FAO/IFA, 2001.

The estimated total fertilizer N loss from all the rice ecologies in Asia is 9.6 Mt/year (1995-97). Thus, 65 percent of applied mineral N (worth US$3 120 million) is lost to the environment, with adverse environmental impacts such as greenhouse effects, diminishing stratospheric ozone, acid rain, changes in the global N cycle and nitrate pollution of surface water and groundwater.

In terms of prevailing crop yields and nitrogen-use efficiency levels, an interesting picture emerges. Most countries, with the exception of the Republic of Korea, Japan and China, fall within the medium- and low-yield groups, indicating considerable scope for raising yields. Of the various N-use efficiency combinations, M and H appear to be the most promising options for minimizing the yield gap while enhancing benefits in terms of dosage economies and reduced environmental threat. Although the level of N use in the Republic of Korea (178 kg/ha) is double that in Japan (88 kg/ha), the yield difference is small (0.2 t/ha) as a result of the efficiency factor. Enhanced N-use efficiency in the Republic of Korea may lead to optimized N use while maintaining yield levels similar to those in Japan. For countries such as China, Indonesia and Iran, improved N-use efficiency accompanied by optimization of N-use levels would be a suitable approach for closing the yield gap. The possibilities for raising yields in Viet Nam, Sri Lanka, Malaysia, India and Pakistan remain high, provided that the prevailing low N-use efficiency levels can be improved. The enhancement of N use to medium levels, coupled with efficiency improvement measures, is important for the Philippines, Bangladesh and Thailand.

Management Practices for Increasing Nitrogen-use Efficiency

The required nutrient management procedures are well understood in general terms. Moreover, methodologies (and associated diagnostics) are emerging to facilitate site-specific nutrient management. FAO promotes an integrated nutrient management approach, which uses manufactured (mineral) fertilizers in appropriate combinations with organic and biological sources.

Nitrogen is the nutrient required in the largest quantities while urea is the principal nitrogenous fertilizer for rice production. However, N from urea is subject to considerable losses to the atmosphere and runoff water in the rice ecosystem, especially where urea is broadcast on standing water. On a global level, more than 55 percent of the N applied through fertilizers to irrigated rice is not taken up. Incorporating urea into the water-saturated soil at a depth of 7 to 10 cm (a reduced zone) minimizes N losses by preventing N from being dissolved in floodwater or oxidized in soil near the surface. This practice is environmentally friendly and enables higher yields with less fertilizer N.

Extensive on-farm trials suggest that the quantification of limiting factors and the adoption of appropriate crop and nutrition management practices can minimize the effects of diminishing returns at increasing N application rates with considerable agronomic, economic and environmental benefits. The (ranked) limiting factors that smallholder rice farmers using prill (or granular) urea can address immediately are: i) too few splits (divided applications of a particular seasonal total fertilizer N) resulting in substantial N losses and consequent inadequate N supply to meet crop requirements at various growth stages; ii) cultivars that may be insufficiently N responsive; and iii) inadequate initial plant population.

Research suggests that farmers and extension agencies can improve N-use efficiency, especially in irrigated rice farming, by using technologies and procedures that can be either incorporated immediately into ongoing national extension programmes or brought rapidly to a state of applicability through relatively small programmes of ecozone-specific adaptive agronomic investigation by extension-cum-research teams.

A multilocation on-farm trial/demonstration project on irrigated rice (1995-98), funded by Japan and implemented by FAO in Indonesia, the Philippines and Malaysia, obtained the following results:

- For a representative grain yield, a total of 58 kg/ha N applied in five equal splits could achieve the same response to that obtained from a seasonal total of 70 kg/ha N applied in three equal splits (a saving of 17%). Leaf-colour chart (for chlorophyll content) guided top-dressing applications showed good promise.
- Deep-placed Indonesian tablet (super granule) urea enabled a 21 percent N saving (compared to 70 kg/ha N applied as prill urea in three splits) and was found to be economically sound.
- Substitution of one-quarter of 70 kg/ha prill urea N by commercial organic fertilizer N resulted in a saving of 8 kg/ha N (11%) of prill urea.
- Banded and paired-row systems of rice establishment, with a corresponding concentration of fertilizer N within these bands and paired rows, resulted in N savings of up to 20 percent (compared to square-array transplanting).
- Establishing an initial rice-plant population density that would result in 40 000 spikelets/m^2 at harvest achieved an N saving of 25 percent as compared to typical suboptimal populations; the option for increased plant population (perhaps by transplanting 4-5 rather than 2-3seedlings per hill, and at 0.20 x 0.20 m rather than at 0.25 x 0.25 m spacing) proved worthwhile, particularly as the influence of diminishing return to N was less than for other technology options.

- A re-analysis (using Mitscherlich procedures) of existing data sets for ecozone-specific N responses of various farmer-preferred rice cultivars could help identify cultivars with high grain yield responses to realistic N applications in particular ecozones.
- At prevailing prices, prill urea (especially with five splits), deep-placed tablet urea and prill urea in combination with plant-derived N (organic fertilizers) were the most cost-effective of the N formulations and types evaluated.
- For the more efficient technologies, grain yield enhancements were typically 30 to 60 percent higher in more N-responsive than in less N-responsive ecozones, and benefit-cost ratios were 12: 4 as compared to 6: 7.
- Technologies identified as appropriate for immediate adoption could be applied in combination. Deep placement of tablet urea in banded or in paired row configurations yielded savings of about 30 percent. However, there was no conclusive information on the practicability and costs of mechanizing such systems. There is a need to determine the N savings that might result where other individual technologies are combined.

Table : Urea deep placement vs broadcast application

Type of trial	*Increment in yield (%)*	*Increment in net revenue (US$/ha)*
Bangladesh		
Farmer trials (boro, 2000)	16[a]	140
Farmer trials (aman, 2000)	20[b]	75
Farmer trials (boro, 2001)	25[c]	148
Result demonstrations (boro, 2001)	20[c]	129
Viet Nam		
Farmer trials (spring paddy, 2001)	14[d]	62
Farmer trials (summer paddy, 2001)	14[e]	78

[a] 77 kg N/ha.

[b] 52 kg N/ha as UDPvs 80 kg N/ha as broadcast prill urea (farmers' practice).

[c] 77 kg N/ha as UDPvs 110 kg N/ha broadcast prill urea (farmers' practice).

[d] N application rates of UDP, on average 35% less than in farmers' practice.

[e] N application rates of UDP, on average 44% less than in farmers' practice.

In Bangladesh, where urea deep placement (UDP) technology is adopted on about 10 percent of the land under rice cultivation, paddy yields have increased by 16 to 25 percent with UDP compared to broadcast urea in three split applications. The additional cost- of making the briquettes (US$10/t of urea) and of labour to place the briquettes (6-8 workdays/ha)-

was about US$13/ha. The increase in net revenue was between US$75 and US$148 per ha. These revenue increments result from the use of less fertilizer and the increase in the per-hectare yield in relation to farmers' practices. The higher yields from UDPwere also due to other improved management practices, such as line planting with optimum space and plant population, better water management and plant protection. In Viet Nam, the consequent yield increase of 14 percent resulted in a net revenue gain of between US$62 and US$78 per ha.

Urea coated with Nimin, a commercial extract from neem (*Azadirechta indica*) seed, has been widely tested, especially in India. This reasonably inexpensive biological product shows great promise for resource-poor farmers, with an average yield increase of between 5 and 10 percent over uncoated prill urea. Super granules made with Nimin-coated urea and placed deep show a slight improvement over the UDP technology.

At national level, the widespread adoption of suitable combinations of more efficient technologies could generate savings of about 30 percent in the total of prill urea applied to rice. This would imply an upgrading of the average N efficiency from 35 to 50 percent. This could generate additional annual savings of about 1.3 to 1.6 Mt (worth about US$500 million). Such efficiency improvements may lead to a reduction in future manufacturing investments of some US$1 500 million, and would save local economies the cost of remedying water pollution. Furthermore, where there are subsidies on urea production and distribution, it would also lessen the burden on the public purse.

Research and Development Issues

The prime concerns relating to rice production are: yield gaps, yield stagnation in irrigated ecology, declining productivity in intensively cultivated areas (localized), low fertilizer efficiency (especially N) and the environmental consequences of input use. Inadequate crop management practices are a major constraint to achieving yield potentials. These inadequacies also lead to inefficient input use. Improved crop management technologies need to be transferred and adapted to local conditions. The current deficiencies are the result of inadequate development and extension. Furthermore, there is a need for strong policy support in fostering farmer-participatory, environmentally-friendly and efficient nutrient management programmes (Roy *et al.*, 2002).

Projections to 2015 and 2030 suggest substantial increases in the rates of N applied to rice, with major environmental implications. The relatively low price of urea N fertilizer (compared to other major nutrients) prevents the widespread adoption of efficient, balanced and environmentally friendly nutrient management. The environmental pricing of urea N, coupled with integrated plant nutrient management, may be an appropriate procedure for fostering N-use efficiency in Asian rice systems.

Nutrient balance studies in many ecosystems have exposed the inadequacies relating to nutrient replenishment and the consequent mining of soil nutrient reserves. These failures need to be rectified through the balanced and efficient use of organic and inorganic plant nutrient sources and through knowledge-intensive integrated soil and nutrient management practices. Compared with input-intensive technologies, such practices are more cost-effective, sustainable and environmentally benign.

Inadequate information on quantified N-uptake efficiency for the various component technologies has been a major constraint to on-farm adoption. Precise quantification is needed for: the major ecozones; varied application rates; farmer-preferred rice cultivars; and transplanted and direct-seeding systems.

Furthermore, there needs to be an assessment of the efficiency (and N application decreases) that would result from combinations of the more efficient individual N technology components. As N application in accordance with the nutrient requirements at the various crop-growth stages is important for fertilizer-use efficiency technologies, colour-card guided top-dressing of N holds good promise. However, such colour cards need to be calibrated for specific ecoregions.

Legume green manures for subsequent rice crops offer considerable potential as partial or total substitutes for N fertilizers, because they can also bring back the lost nutrients from the subsoil to the surface. However, the adoption of green manuring technology has been poor in many areas (especially in countries with limited land resources and high subsidies on mineral N fertilizers), because farmers perceive it to be laborious and uneconomical. As farmers look for multiple benefits from their investments, green manure systems will need to provide additional benefits, such as food, fodder, fuel and commercial products.

The induction of nitrogen-fixing capabilities in rice through bioengineering processes, such as induced symbiosis and Nif-gene transfer, offers significant potential for reducing the need for mineral N applications. Furthermore, free-living blue green algae and Azolla with its symbiont *Anabaena azollae* are principal biological nitrogen fixation (BNF) agents in rice fields. Evidence suggests that algal inoculations can yield a benefit equivalent to the application of 25 to 30 kg/ha N.

However, their adoption has been constrained by several factors, including maintenance of Azolla inocula between cropping seasons. Measures, such as the selection of superior germplasm, the development of improved Azolla hybrids, and efforts to induce sporulation, should enhance the adoption of this technology.

Conclusions

Growing concerns about the environmental consequences of mineral N use and its future costs have highlighted three approaches:

- reduce N losses by adopting efficient fertilizer use technologies;
- recycle animal and plant residues; and
- develop BNF within the framework of an integrated nutrient management approach.

Even with reduced reliance on mineral N, a firm commitment and the provision of adequate resources from all stakeholders (governments, fertilizer sector, NGOs and international agencies) in transferring improved sound N management technologies as part of a bottom-up participatory approach, can produce more rice, increase farmers' incomes and reduce environmental pollution.

Azotobacter Biofertilizers Technology and Quality Control

Azotobacter Vinelandii is a free-living bacterium that can fix atmosferic nitrogen into the soil, being a great source to obtain a natural biofertilizer that can be used in the cultivation of most crops.

It is a great source of nitrogen to meet the needs of crops because also has the capabilities to cause a rejuvenation of soil microbiology to tap out the biological fixation of nitrogen. Benefits:

- It improves seed germination and plant growth
- Azotobacter are tolerant to high salts.
- It can benefit crops by Nitrogen fixation, growth promoting substances, fungi static substances.
- Azotobacter is heaviest breathing organism and requires a large amount of organic carbon for its growth.
- It is poor competitor for nutrients in soil and hence its growth promoting substances, fungistatic substances.
- It thrives even in alkaline soils.
- Azotobacter is less effective in soils with poor organic matter content.

Characteristics of Azotobacter

Azotobacter is Gram negative bacteria, polymorphic i.e. they are of different sizes and shapes. Their size ranges from 2-10x1-2.5 m., young cell possess peritrichous flegella and are used as locomotive organs. Old population of bacteria includes encapsulated forms and have enhanced resistant to heat, desication and adverse conditions.

The cyst germinates under favourable conditions to give vegetative cells. They also produce polysachharides. Azotobacter spp.,are sensitive to acidic pH, high salts, and temperature above 35^{0}C.

There are four important species of Azotobacter viz. A.Chroococcum, A.agilis, A.paspali and A.vinelandii of which A.chroococcum is most commonly found in our soils.

Nitrogen fixation by *Azotobacter*: The species of Azotobacter are known to fix on an average 10 mg.of N/g of sugar in pure culture on a nitrogen free medium. A maximum of 30 mg. N fixed per gram of sugar was reported by lopatina. However, Azotobacter is a poor competitor for nutrients in soil. Most efficient strains of Azotobacter would need to oxidise about 1000 kg of organic matter for fixing 30 kg of N/ha. This does not sound realistic for our soils which have very low active carbon status. Besides, soil is inhabitated by a large variety of otherr microbes, all of which compete for the active carbon.

Azotobacter in Soil

Soils are habited by a very large number of microbial species. The co-existence of the relative populations of each one of the species is determined by ecological factors prevailing in the soil. These various species survive in soil while maintaining a balance of population is between various microbial species within certain limits.

A normal population of Azotobacter could be about than 10 thousand to 1 lakh/g of soil and it is mostly influenced by other micro-organisms present in soil. There are some micro-organism which stimulate the Azotobacter population in soil thereby increasing the nitrogen fixation by Azotobacter. On the other hand there are some micro-organisms which adversely affect the Azotobacter population and hence nitrogen fixation process is hampered. For example cephallosporium is most commonly found organisms in soil which restricts the growth of Azotobacter.

Additional Benefits: Azotobacter also produces some substances which check the plant pathogens such as Alternaria, Fusarium and Helminthosporium. Hence Azotobacter also acts as a biological control agent.

Functions of Azotobacter: Azotobacter naturally fixes atmospheric nitrogen in the rhizosphere. There are different strains of Azotobacter each has varied chemical, biological and other characters. However, some strains have higher nitrogen fixing ability than others.. Azotobacter uses carbon for its metabolism from simple or compound substances of carbonaceous in nature. Besides carbon, Azotobacter also requires calcium for nitrogen fixation. Similarly, a medium used for growth of Azotobacter is required to have presence of organic nitrogen, micro-nutrients and salt in order to enhance the nitrogen fixing ability of Azotobacter.

Azotobacter also produces fixation of Thiomin, Riboflavin, Nicotin, indol acitic acid and giberalin. When Azotobacter is applied to seeds, seed germination is improved to a considerable extent, so also it controls plant diseases due to above substances produced by Azotobacter.

Selection of Azotobacter strains: After isolation of Azotobacter from soil its purity is tested in the laboratory in a pure form. In fertile soil spp.A.Chroococcum is found, commonly. The organism is aerobic in nature, requires oxygen for its growth. In old culture Malinin chemical is formed which gives the blackish colour to the culture. The organism is prominently found in alkaline or neutral soils. Strains of Azotobacters vary in their nitrogen fixing ability which depends upon pH of soil, crop and atmosphere of soil. Therefore nitrogen fixing capacity of strains is tested frequently.

(Azotobacter) (Bio Fertilizers) Chemicals

This is an another effective nitrogen fixing bacterium. It also fixes atmospheric nitrogen into the soil. *azotobacter* species are gram-negative, aerobic soil-dwelling bacteria.

e specially cereals, vegetables, fruits, trees, sugarcane, cotton, grapes, banana, etc. Are known to get addition nitrogen requirements from azotobacter. Beneficial aspect:

- reduces expenditure on chemical fertilizers up to 25%
- it helps for better crop growth and seedling establishment.
- increase crop yield over 25%
- increases soil microbial activity
- azotobacter also increases germination of seeds
- enhances overall soil fertility
- produces various growth promoting substances, which ultimately increase seed germination, plant stand and pant vigor.

a. Growth of Azotobacter : Usually Azotobacter is grown on a solid medium free of nitrogen. After some times (6 months) old growth of Azotobacter is transferred to a fresh solid medium to renew the growth. This procedure is repeated periodically so that the culture can be maintained in good condition.

b. Production:

i. Mother culture: A pure growth of any organism on a small scale is called as a mother culture. Mother culture is always prepared in a conical flask of 500 or 1000 ml. Capacity and then this mother culture is used for further production.

For this purpose, one litre conical flasks are taken to

which 500 ml of broth of nitrogen free medium is added and these flasks are then plugged with non-absorbent cotton, sterilized in an auto slave for 15-20 minutes at 75 lbs pressure for 15 minutes. Flasks are then inoculated with mother culture with the help of inoculating needle aseptically. The flasks are transferred to shaker and shaking is done for 72-90 hours so as to get optimum growth of bacteria in broth. Bacteria are multiplied by binary method i.e. cell division. After about 90 days, the number of per milliliters comes to about 100 crores. Total growth of bacteria in this broth means starter culture or mother culture, which should carefully be done, since further purity of biofertilizer or quality of biofertilizer depends upon how mother culture is prepared.

ii. Production on a large scale: Azotobacter is multiplied on a large scale by two ways viz. Fermenter and Shaker. The fermenter is most automatic and accurate method of multiplication of any micro-organism. In this method, the medium is taken in a fermenter and then sterilized. After this pH of the medium is adjusted and 1% mother culture is added. In order to get an optimum growth of the Azotobacter required temperature and oxygen supply is adjusted so that concentrated broth is made. This concentrated broth of the culture is then mixed with a carrier previously sterilized and bio-fertilizers is prepared. Depending upon the demand and supply suitable fermenter is selected.

In the 2nd method i.e. shake method, a suitable medium is prepared transferred to conical flask of suitable capacity. These flasks are then sterilized in an autoclave at 15 lbs pressure for 15 minutes. Each flasks is inoclulated with 10 ml mother culture and they are transferred to shaker for multiplication where they are kept for 72-90 hours. This broth is mixed with a suitable carrier previously sterilized. Thus biofertilizer is prepared, filled in plastic bags and stored in cool place.

Selection of Carrier

A carrier is nothing but a substance which has high organic matter, higher water holding capacity and supports the growth of organism. In order to transport the biofertilizer and becomes easy to use the suitable carrier is selected. Generally Lignite cool, compost and peat soil are suitable carriers for Azotobacter. Out of these carriers lignite is most suitable for this organism, since it is cheaper, keeps organism living for longer period and does not lower the quality of bio-fertilizers.

The lignite comes in clouds and hence it is ground in fine powder by grinding machine. Its finesses should be 250-300 mesh. The pH of the carrier is adjusted to neutral by adding $CaCO_3$. The lignite naturally has a variety of micro-organism and hence it is sterilized in autoclave at 30 lbs. Pressure for 30 minutes. After this the broth is mixed with lignite 1:2 proportion by following method.

Galvanized trays are sterilized and used. To these trays, previously sterilized lignite is transferred and broth is then added (lignite2: broth 1) and mixed properly. Trays are then kept one above the other for 10-12 hours for allowing the organism to multiply in the carrier. This mixture is then filled in plastic bags of 250 g or 500 g capacity. Plastic bags are properly. Trays are then kept one above the other for 10-12 hours for allowing the organism to multiply in the carrier. This mixture is then filled in plastic bags of 250 g or 500 g capacity. Plastic bags are properly sealed. All the required information such as name of biofertilizer, method of use expiry date, etc. is printed on plastic bags. In this way biofertilizer is ready to sell or use. If biofertilizer is used immediately then bags are stored in cool place otherwise they should be stored in cold storage in order to keep biofertilizer in good quality.

As per ISI standards, one gram of biofertilizer immediately after it is prepared should have one crore cells of bacteria and 15 days before expiry date one gram of biofertilizer should have 10 lakh bacteria. If biofertilizer is stored at 15-20 ^{0}C then it will remain effective for 6 months. However, at 0 to 4 ^{0}C (cold storage) the bacteria will remain active for 2 years. The storage periods are decided after testing the biofertilizer for that particular storage conditions, such temperature and humidity.

Use of Biofertilizer

A plant needs nitrogen for its growth and Azotobacter fixes atmospheric nitrogen non-symbiotically. Therefore, all plants, trees, vegetables, get benefited. However, especially cereals, vegetables, fruits, trees, sugarcane, cotton, grapes, banana, etc. are known to get addition nitrogen requirements from Azotobacter. Azotobacter also increases germination of seeds. Seeds having less germinating percent if inoculated can increase germination by 20-30%.

How to Apply Bio-fertilizer ?

a. *Seed inoculation:* On the basis of efficiency of Azotobacter, other micro-organisms present in the soil, benefits obtained from biofertilizer and expenditure it has been fixed to use Azotobacter-bio-fertilizer at the rate of 250 g biofertilizer for 10-15 kg. If one knows this proportion then take a definite quantity of seed to be inoculated. The required quantity of fresh biofertilizer is secured and a slurry is made by adding adequate, quantity of water. This slurry is uniformly applied to seed, seed is then dried in shed and sown. Some stickers are used in order to adher biofertilizer to seeds. Viz. Jaggery or gum arebia.

b. *Seedling inoculation:* This method of inoculation is used where seedlings are used to grow the crop. In this method, seedlings required for one acre are inoculated using 4-5 packets (2-2.5 kg). For this, in a bucket adequate quantity of water is taken and biofertilizer from these packets is added to bucket and mixed properly. Roots or seedlings are then dipped in this mixture so as to enable roots to get inoculum. These seedlings are then transplanted e.g. Tomato, Rice, Onion, Cole, Crops, flowers.

c. *Self inoculation or tubez inoculation:* In this method 50 litres of water is taken in a drum and 4-5 kg of Azotobacter biofertilizer is added and mixed properly. Sets are required for one acre of land are dipped in this mixture. Potato tubers are dipped in the mixture of biofertilizer and planting is done.

d. *Soil application:* This method is mostly used for fruit crops, sugarcane, and trees. At the time of planting fruit tree 20 g of biofertilizer mixed with compost is to be added per sappling, when trees became matured the same quantity of biofertilizer is applied.

In sugarcane after two to three months of planting i.e. before earthing up 5-6 kg of biofertilizer per acre is applied by mixing with compost or soil. Although, Azotobacter fixes nitrogen non-symbiotically, it also fixes atmospheric nitrogen in the rhizospere region i.e. soil around the seedlings or trees. Biofertilizer applied to seed or seedlings bacteria remain around seeds or seedlings and use organic carbon for their metabolism. When seeds are germinated or seedlings set in soil they leave or exude root exudates which become food of these bacteria. They grow on these substances which include sugars, organic acids, amino acids and fix atmospheric nitrogen most efficiently. Nitrogen so fixed by these bacteria becomes available to plants after dead and degradation of bacterial cells.

Conclusions

1. Azotobacter contributes moderate benefits
2. Azotobacter is heaviest breathing organism and requires a large amount of organic carbon for its growth.
3. It is poor competitor for nutrients in soil and hence its growth promoting substances, fungistatic substances.
4. It can benefit crops by Nitrogen fixation, growth promoting substances, fungi static substances.

5. Azotobacter is less effective in soils with poor organic matter content.
6. It improves seed germination and plant growth
7. It thrives even in alkaline soils.

Azospirillum Biofertilizer Technology and Quality Control

The ability to produce enough food for a country is an essential task of any society. If a nation must import food they will always be at the mercy of other nations for their very existance. The basis of any good agricultural system is the ability to produce large amounts of crops. The crops go not only to feed people but in large portion to feeding, and fattening cattle.

Plants need light, water, carbon dioxide, and nitrogen to grow. Nitrogen is undoubtly the most important plant nutrient. That is why the production of inorganic nitrogen fertilizers has been such a benefit to farmers in developed nations. Unfortunately, these fertilizers are expensive, tend to leach from soils, and eventually contaminate our drinking water. Developing nations can not afford to fertilize their fields in the amounts needed year after year to get the yields needed. Developed nations can not afford the continuing pollution of surface waters.

How can this problem be solved? Just ask a soil microbiologist and they can give you several alternative solutions to the application of inorganic fertilizers. One such solution is the application of bacteria called *Azospirillum*. This group of bacteria can take nitrogen from the air and provide it to the plant (nitrogen fixation).

Nitrogen fixation is not the major benefit *Azospirillum* can give to plants. When *Azospirillum* are placed on the roots of plants they have the ablity to increase the numbers of root hairs (from Iowa State University) on each root. These root hairs are important in helping the plant get water from the soil. This benefit the plant derives from its association with *Azospirillum* helps the plants do better in drier places.

Another benefit the plant gets from associating with *Azospirillum* is that the bacteria produce plant growth factors (auxins) that cause the plant to produce more roots. The better the root system, the stronger the plant. Stronger plants produce more food. The plants given *Azospirillum* are also as productive as heavily fertilized plants that do not have the *Azospirillum* added.

Researchers have found that they can apply less fertilizer per acre and get good yields. This would help farmers in developing countries increase their yields without significantly increasing their costs. Using less fertilizer would also lower the amount of fertilizer polluting the drinking waters in developed countries.

Unfortunately, the yields have been somewhat inconsistent. Some portions of a farmer's land may respond very well to the addition of *Azospirillum*.

Other fields may not respond at all. These variations of response to the addition of *Azospirillum* are due to the wide variations in soil, plant and microfloral (other bacteria in the soil) components.

Azospirillum Culture

The cells of Azospirillum remain in association with the roots and fix part of the atmospheric nitrogen.

Azospirillum cultures are useful for the cereals and cash crops viz. Wheat, Paddy, Bajra, Jowar, Maize, Mustard, Cotton, Cumin, Banana, Sugarcane, Tobacco, Castor, Vegetables etc., as well as horticultural crops.

Advantages

- The effective strain used in Azotobacter culture fixes about 15 – 20 Kg. atmospheric nitrogen per hectare.
- Certain growth promoting substances released by these cultures are useful for increasing the seed germination, plant growth and ultimately the yield.
- In certain condition they also exhibit anti-fungal activities and thereby fungal diseases may be controlled indirectly.
- About 10 to 15% increase of crop yield can be achieved with the use of these cultures.

9

Rhizobial Biofertilizer for Leguminous Crops

Rhizobia are soil bacteria that fix nitrogen (diazotrophy) after becoming established inside root nodules of legumes (Fabaceae). Rhizobia require a plant host; they cannot independently fix nitrogen. Morphologically, they are generally gram negative, motile, non-sporulating rods.

History

The first species of rhizobia, *R. leguminosarum*, was identified in 1889, and all further species were initially placed in the *Rhizobium* genus. However, more advanced methods of analysis have revised this classification, and now there are many in other genera. Most research has been done on crop and forage legumes such as clover, alfalfa, beans, and soy; recently, more work is occurring on North American legumes.

The word *rhizobia* comes from the Ancient Greek, "rhíza," meaning "root," "bios," meaning "life." The word *rhizobium* is still sometimes used as the singular form of *rhizobia*.

Taxonomy

Rhizobia are a paraphyletic group which fall into two classes of the proteobacteria—the alpha- and beta-proteobacteria. As shown below, most belong to the order Rhizobiales but several rhizobia occur in distinct bacterial orders of the proteobacteria.

These groups include a variety of non-symbiotic bacteria. For instance, the plant pathogen *Agrobacterium* is a closer relative of *Rhizobium* than the *Bradyrhizobium* that nodulate soybean (and may not really be a separate genus).

The genes responsible for the symbiosis with plants, however, may be more closely related than the organisms themselves, acquired by horizontal transfer (via bacterial conjugation) rather than vertical gene transfer (from a common ancestor).

Importance in Agriculture

Although much of the nitrogen is removed when protein-rich grain or hay is harvested, significant amounts can remain in the soil for future crops. This is especially important when nitrogen fertilizer is not used, as in organic rotation schemes or some less-industrialized countries. Nitrogen is the most commonly deficient nutrient in many soils around the world and it is the most commonly supplied plant nutrient. Supply of nitrogen through fertilizers has severe environmental concerns.

Symbiosis

Rhizobia are unique because they live in a symbiotic relationship with legumes. Common crop and forage legumes are peas, beans, clover, and soy.

Infection and Signal Exchange

The symbiotic relationship implies a signal exchange between both partners that leads to mutual recognition and development of symbiotic structures. Rhizobia live in the soil where they are able to sense flavonoids secreted by the root of their host legume plant. Flavonoids trigger the secretion of Nod factors, which in turn are recognized by the host plant and can lead to root hair deformation and several cellular responses such as ion fluxes. The best known infection mechanism is called intracellular infection, in this case the rhizobia enter through a deformed root hair in a similar way to endocytosis, forming an intracellular tube called the infection thread. A second mechanism is called "crack entry", in this case no root hair deformation is observed and the bacteria penetrate between cells, through cracks produced by lateral root emergence. Later on bacteria become intracellular and an infection thread is formed like in intracellular infections.

The infection triggers cell division in the cortex of the root where a new organ, the nodule appears as a result of successive processes.

Nodule Formation and Functioning

Infection threads grow to the nodule, infect its central tissue and release the rhizobia in these cells where they differentiate morphologically into bacteroids and fix nitrogen from the atmosphere into a plant usable form, ammonium (NH_4^+), utilizing the enzyme nitrogenase. In return the plant supplies the bacteria with carbohydrates, proteins, and sufficient oxygen so as not to interfere with the fixation process. Leghaemoglobins, plant proteins similar to human hemoglobins help to provide oxygen for respiration while keeping the free oxygen concentration low enough not to inhibit nitrogenase activity. Recently, it was discovered that a *Bradyrhizobium* strain forms nodules in *Aeschynomene* without producing Nod factors, suggesting the existence of alternative communication signals other than Nod factors.

The legume–rhizobium symbiosis is a classic example of mutualism—rhizobia supply ammonia or amino acids to the plant and in return receive organic acids (principally as the dicarboxylic acids malate and succinate) as a carbon and energy source—but its evolutionary persistence is actually somewhat surprising. Because several unrelated strains infect each individual plant, any one strain could redirect resources from nitrogen fixation to its own reproduction without killing the host plant upon which they all depend. But this form of cheating should be equally tempting for all strains, a classic tragedy of the commons. There are two competing hypotheses for the mechanism that maintains legume-rhizobium symbiosis (though both may occur in nature).

The sanctions hypothesis suggests that plants police cheating rhizobia. Sanctions could take the form of reduced nodule growth, early nodule death, decreased carbon supply to nodules, or reduced oxygen supply to nodules that fix less nitrogen. The partner choice hypothesis proposes that the plant uses pre-nodulation signals from the rhizobia to decide whether to allow nodulation and chooses only non-cheating rhizobia. There is evidence for sanctions in soybean plants, which reduce rhizobium reproduction (perhaps by limiting oxygen supply) in nodules that fix less nitrogen. Similarly, wild lupine plants allocate less resources to nodules containing less-beneficial rhizobia, limiting rhizobial reproduction inside. This is consistent with the definition of sanctions just given, although called "partner choice" by the authors. However, other studies have found no evidence of plant sanctions and instead support the partner choice hypothesis.

Other Diazotrophs

Many other species of bacteria are able to fix nitrogen (diazotrophs), but few are able to associate intimately with plants and colonize specific structures like Legume nodules. Bacteria that do associate with plants include the actinobacteria *Frankia,* which form symbiotic root nodules in actinorhizal plants, and several cyanobacteria (*Nostoc*) associated with aquatic ferns, *Cycas* and *Gunneras*. Free-living diazotrophs are often found in the rhizosphere and in the intercellular spaces of several plants including rice and sugarcane, but in this case the lack of a specialized structure results in poor nutrient transfer efficiency compared to legume or actinorhizal nodules.

Maturity Mapping of the Rhizobial Biofertilizer

After reviewing the literatures related to rhizobial biofertilizer technology and asking two rhizobial experts' opinion, it could be concluded that there were five categories of major performance parameters which would be used to determine the performance of rhizobial biofertilizer technology described in the patents. These five groups of major performance parameters were yield, stability, convenience of use, adaptability and cost, which were described in detail as follows:

Yield The output of legume crops (kg/rai, kg/hec) after inoculating with rhizobial biofertilizers described in the patents. It can be tested by using greenhouse or field trial tests. Yield resulted from two subperformance parameters which were nitrogen-fixing ability and/or hydrogen uptake activity and nodulation and/or competitiveness with indigenous rhizobia. The definitions of these sub-performance parameters were described in the followings:

Nitrogen-fixing ability: Ability of inoculated rhizobial biofertilizers to fix atmospheric nitrogen (N2) and convert into nitrogeneous compounds (ammonia-NH3) which can be directly used by legume crops. The ability could be enhanced by modifying nitrogenase enzyme activities, their cofactors and genes (fix, nif). It can be tested by using acetylene reduction assay.

Hydrogen uptake activity: Ability of inoculated rhizobial biofertilizers to recover hydrogen gas (H2) lost during nitrogen fixation process by converting into proton (H+) and supplying it into the electron transport system to produce energy (ATP) used for nitrogen fixation process. The ability could be enhanced by modifying hydrogenase enzyme activities and their genes (Hup). It can be tested by observations of differences in rates of reduction of methylene blue.

Nodulation : Ability of inoculated rhizobial biofertilizers to form root nodules in the legume crops. It can be tested by using seedling agar-tube method to determine nodule numbers or nodule occupancy per plant.

Competitiveness with indigenous rhizobia: Ability of inoculated rhizobial biofertilizers to compete with indigenous rhizobia already living in the soil to infect root hair and form root nodules in the legume crops.

Stability Ability of rhizobial biofertilizers to be resistant to the adverse conditions and substances, and maintain viability of rhizobia before and after inoculation to the legume crops. Stability consisted of four subperformance parameters which were shelf-life, temperature-resistance, pHresistance and other adverse conditions-resistance. The definitions of these sub-performance parameters were described in the followings:

Shelf-life (storage): The amount of storage time keeping at room temperature (25 oC) since rhizobial biofertilizer production until prior to use with legume crops. During this storage time, the rhizobial biofertilizers still maintain viability of rhizobia and their efficiencies.

Temperature-resistance: Ability of rhizobial biofertilizers to resist to the extreme temperatures such as heat and cold temperature.

pH-resistance: Ability of rhizobial biofertilizers to resist to the extreme pH in the soil such as acidic and basic soil.

Other adverse conditions-resistance: Ability of rhizobial biofertilizers to resist to other adverse conditions and substances mentioned above such as drought, herbicides, fungicides and etc.

Convenience of use Ease of handling and applying rhizobial biofertilizers to inoculate the legume crops. It might come from the form of rhizobial biofertilizers (liquid, powder, granule) and their inoculation methods.

Adaptability Versatility of rhizobial biofertilizers to use with variety of legume crops and soil types. Adaptability consisted of two subperformance parameters which were the number of the legume crops (e.g. pea, soybean, mungbean, clover, alfalfa, vetch) and soil types with which rhizobial biofertilizers can be used.

Cost : The cost of raw materials, substances, instruments and methods for the production of rhizobial biofertilizers.

Microbial Technologies in Human Welfare

Microorganisms and their biosynthetic capabilities have solved many difficult problems in our daily life for human welfare. The various ways in which microorganisms have helped mankind include medical technology, human and animal health, food processing, food safety and quality, genetic engineering, environmental protection, agricultural biotechnology and more effective treatment of agriculture, municipal and industrial waste that provide a most impressive record of achievement. Most of these technological advancements would not have been possible using conventional straight forward chemical and physical engineering methods, or if they were, they would not have been practically or economically feasible and may cause environmental problems. In recent years, microbial technologies have been applied to several agricultural and environmental problems with considerable success. Since microorganisms are useful in eliminating problems associated with the use of domestic and industrial pollutants, they are now widely applied in nature farming and organic agriculture.

Biotransformation

Biotransformation is the chemical modification (or modifications) made by an organism on a chemical compound. If this modification ends in mineral compounds like CO_2, NH_4^+ or H_2O, the biotransformation is called mineralisation. Biotransformation means chemical alteration of chemicals such as (but not limited to) nutrients, amino acids, toxins, or drugs in the body. It is also needed to render nonpolar compounds polar so that they are not reabsorbed in renal tubules and are excreted.

Microbial life is amazingly diverse and microorganisms literally cover the planet.

An average gram of soil contains approximately one billion (1,000,000,000) microbes representing probably several thousand species. Microorganisms have special impact on the whole biosphere. They are the backbone of ecosystems of the zones where light cannot approach. In such zones, chemosynthetic bacteria are present which provide energy and carbon to the other organisms there. Some microbes are decomposers which have ability to recycle the nutrients. Microbes have a special role in biogeochemical cycles. Microbes, especially bacteria, are of great importance because their symbiotic relationship (either positive or negative) have special effects on the ecosystem.

Microorganisms are used for *in-situ* microbial biodegradation or bioremediation of domestic, agricultural and industrial wastes and subsurface pollution in soils, sediments and marine environments. The ability of each microorganism to degrade toxic waste depends on the nature of each contaminant. Since most sites typically have multiple pollutant types, the most effective approach to microbial biodegradation is to use a mixture of bacterial species and strains, each specific to the biodegradation of one or more types of contaminants. It is vital to monitor the composition of the indigenous and added bacteria in order to evaluate the activity level and to permit modifications of the nutrients and other conditions for optimizing the bioremediation process.

Oil Biodegradation

Petroleum oil is toxic, and pollution of the environment by oil causes major ecological concern. Oil spills of coastal regions and the open sea are poorly containable and mitigation is difficult; much of the oil can, however, be eliminated by the hydrocarbon-degrading activities of microbial communities, in particular the hydrocarbonoclastic bacteria (HCB). These organisms can help remedy the ecological damage caused by oil pollution of marine habitats. HCB also have potential biotechnological applications in the areas of bioplastics and biocatalysis.

Waste Biotreatment

Biotreatment, the processing of wastes using living organisms, is an environmentally friendly alternative to other options for treating waste material. Bioreactors have been designed to overcome the various limiting factors of biotreatment processes in highly controlled systems. This versatility in the design of bioreactors allows the treatment of a wide range of wastes under optimized conditions. It is vital to consider various microorganisms and a great number of analyses are often required.

Wastewater Treatment

Wastewater treatment processes are geared towards one purpose: cleaning up water. Recent application of molecular techniques is unveiling the microbial composition and architecture of the complex communities involved in the treatment processes. It is now recognized that wastewater processes harbor a vast variety of microorganisms most of which are yet-to-be cultured, hence uncharacterized. Metagenomic technology is being used to study the diversity, structure and functions of microbial communities in nitrifying processes, anaerobic ammonia oxidation processes and methane fermenting processes.

Environmental Genomics of Cyanobacteria

The application of molecular biology and genomics to environmental microbiology has led to the discovery of a huge complexity in natural communities of microbes. Diversity surveying, community fingerprinting and functional interrogation of natural populations have become common, enabled by a range of molecular and bioinformatics techniques. Recent studies on the ecology of Cyanobacteria have covered many habitats and have demonstrated that cyanobacterial communities tend to be habitat-specific and that much genetic diversity is concealed among morphologically simple types. Molecular, bioinformatics, physiological and geochemical techniques have combined in the study of natural communities of these bacteria.

Corynebacteria

Corynebacteria are a diverse group Gram-positive bacteria found in a range of different ecological niches such as soil, vegetables, sewage, skin, and cheese smear. Some, such as *Corynebacterium diphtheriae*, are important pathogens while others, such as *Corynebacterium glutamicum*, are of immense industrial importance. *C. glutamicum* is one of the biotechnologically most important bacterial species with an annual production of more than two million tons of amino acids, mainly L-glutamate and L-lysine.

Legionella

Legionella is common in many environments, with at least 50 species and 70 serogroups identified. *Legionella* is commonly found in aquatic habitats where its ability to survive and to multiply within different protozoa equips the bacterium to be transmissible and pathogenic to humans.

Archaea

Originally, Archaea were once thought of as extremophiles existing only in hostile environments but have since been found in all habitats and may contribute up to 20% of total biomass. Archaea are particularly common in the oceans, and the archaea in plankton may be one of the most abundant groups of organisms on the planet. Archaea are subdivided into four phyla of which two, the Crenarchaeota and the Euryarchaeota, are most intensively studied.

Lactobacillus

Lactobacillus species are found in the environment mainly associated with plant material. They are also found in the gastrointestinal tract of humans, where they are symbiotic and make up a portion of the gut flora.

Aspergillus

Aspergillus spores are common components of aerosols where they drift on air currents, dispersing themselves both short and long distances depending on environmental conditions. When the spores come in contact with a solid or liquid surface, they are deposited and if conditions of moisture are right, they germinate. The ability to disperse globally in air currents and to grow almost anywhere when appropriate food and water are available means that ubiquitous is among the most common adjectives used to describe these moulds.

Rhizobia

Symbiotic nitrogen fixation is a mutualistic process in which bacteria reside inside plants and reduce atmospheric nitrogen to ammonia. This ammonia can then be used by the plant for the synthesis of proteins and other nitrogen-containing compounds such as nucleic acids. The Gram-negative soil bacteria that carry out this process are collectively referred to as rhizobia (from the Greek words Riza = Root and Bios = Life). The process of symbiotic nitrogen fixation is of agricultural and ecological significance because plants capable of nitrogen fixation do not need to compete for limited quantities of soil nitrogen, nor do they require expensive nitrogenous fertilizers that can be harmful to the environment.

Microalgae

Algae are a highly diverse group of protists, ranging from simple, unicellular organisms to complex, multicellular entities with a range of differentiated tissues and distinct organs. They are found among diverse aquatic ecosystems and play important roles by supplying carbon and energy as well as providing habitat to other members of the biological communities. Some algae cause significant environmental and health problems. There are three algal groups: the dinoflagellates, the diatoms and the haptophytes. their 3 main phylla are chlorophyta,rhodophyta and phaeophyta.

Anaerobic Protozoa

Diplomonads are a group of mitochondrion-lacking, binucleated flagellates found in anaerobic or micro-aerophilic environments. Most research on diplomonads has focused on *Giardia,* which is a major cause of water-borne enteric disease in humans and other animals. The first diplomonad to have its genome sequenced was a *Giardia* isolate (WB) and the 11.7 million basepair genome is compact in structure and content with simplified basic cellular machineries and metabolism.

Water Microbiology

An adequate supply of safe drinking water is one of the major prerequisites for a healthy life, but waterborne diseases are still a major cause of death in many parts of the world, particularly in young children, the elderly, or those with compromised immune systems. As the epidemiology of waterborne diseases is changing, there is a growing global public health concern about new and reemerging infectious diseases that are occurring through a complex interaction of social, economic, evolutionary, and ecological factors. An important challenge is therefore the rapid, specific and sensitive detection of waterborne pathogens. Presently, microbial tests are based essentially on time-consuming culture methods. However, newer enzymatic, immunological and genetic methods are being developed to replace and/or support classical approaches to microbial detection. Moreover, innovations in nanotechnology and nanosciences are having a significant impact in biodiagnostics, where a number of nanoparticle-based assays and nanodevices have been introduced for biomolecular detection.

Molecular techniques based on genomics, proteomics and transcriptomics are rapidly growing as complete microbial genome sequences are becoming available, and advances are made in sequencing technology, analytical biochemistry, microfluidics and data analysis. While the clinical and food industries are increasingly adapting these techniques, there appear to be major challenges in detecting health-related microbes in source and treated drinking waters. This is due in part to the low density of pathogens in water, necessitating significant processing of large volume samples. From the vast panorama of available molecular techniques, some are finding a place in the water industry: Quantitative PCR, protein detection and immunological approaches, loop-mediated isothermal amplification (LAMP), microarrays.

Sensory Mechanisms

Bacteria have evolved abilities to regulate aspects of their behaviour (such as gene expression) in response to signals in the intracellular and extracellular environment. The interaction of a signal with its receptor (usually a protein or RNA molecule) triggers a series of events that lead to reprogramming of cellular physiology, typically as a consequence of altered patterns of gene expression. In this way, the bacterial cell is able to mount appropriate and effective responses to changing physical and/ or chemical environments. The versatility with which many bacteria adapt to environmental change underlies many important aspects of microbiology. For example, pathogens encounter multiple environments as they invade a host from the outside, and then progress through different sites within host tissues.

There is growing evidence that pathogenic bacteria make use of physical and chemical cues to signal their presence in a suitable host, and need to adapt to the host environment in order to mount a successful infection. On the other hand, it should not be assumed that all signals to which bacteria must respond originate in the extracellular environment. For many species, even the cosseted life in a laboratory shake flask is 'stressful', in the sense that there is often a need to avoid or reverse the effects of harmful intermediates or by-products of metabolism. For example, all organisms that use dioxygen as a terminal electron acceptor have to deal with the reactive oxygen species that arise as adventitious by-products of aerobic metabolism. In bacteria, multiple protein receptors for oxygen radicals have been described, which control the expression of genes encoding enzymes that detoxify oxygen radicals or repair the damage that they cause.

Drug Metabolism

The metabolism of a drug or toxin in a body is an example of a biotransformation. Typically the body deals with a foreign compound by making it more soluble, to increase the rate of its excretion through the urine. There are a number of different process that can occur; the pathways of drug metabolism can be divided into:

- phase I
- phase II.

Drugs can undergo one of four potential biotransformations: Active Drug to Inactive Metabolite, Active Drug to Active Metabolite, Inactive Drug to Active Metabolite, Active Drug to Toxic Metabolite (biotoxification).

Phase I Reaction

- Includes oxidative, reductive and hydrolytic reactions.
- In these type of reactions, a polar group is either introduced or unmasked, so the drug molecule becomes more water-soluble and can be excreted.
- Reactions are non-synthetic in nature & generally produce a more water soluble & less active metabolites.
- The majority of metabolites are generated by a common hydroxylating enzyme system known as Cytochrome P450.

Phase II Reaction

- These reactions involve covalent attachment of small polar endogenous molecule such as glucuronic acid, sulfate, or glycine to form water-soluble compounds.
- This is also known as a *conjugation reaction*.
- The final compounds have a larger molecular weight.

Microbial Biotransformation

Biotransformation of various pollutants is a sustainable way to clean up contaminated environments. These bioremediation and biotransformation methods harness the naturally occurring, microbial catabolic diversity to degrade, transform or accumulate a huge range of compounds including hydrocarbons (e.g. oil), polychlorinated biphenyls (PCBs), polyaromatic hydrocarbons (PAHs), pharmaceutical substances, radionuclides and metals. Major methodological breakthroughs in recent years have enabled detailed genomic, metagenomic, proteomic, bioinformatic and other high-throughput analyses of environmentally relevant microorganisms providing unprecedented insights into biotransformation and biodegradative pathways and the ability of organisms to adapt to changing environmental conditions.

Biological processes play a major role in the removal of contaminants and pollutants from the environment. Some microorganisms possess an astonishing catabolic versatility to degrade or transform such compounds. New methodological breakthroughs in sequencing, genomics, proteomics, bioinformatics and imaging are producing vast amounts of information. In the field of Environmental Microbiology, genome-based global studies open a new era providing unprecedented *in silico* views of metabolic and regulatory networks, as well as clues to the evolution of biochemical pathways relevant to biotransformation and to the molecular adaptation strategies to changing environmental conditions. Functional genomic and metagenomic approaches are increasing our understanding of the relative importance of different pathways and regulatory networks to carbon flux in particular environments and for particular compounds and they are accelerating the development of bioremediation technologies and biotransformation processes. Also there is other approach of biotransformation called enzymatic biotransformation.

Oil Biodegradation

Petroleum oil is toxic for most life forms and episodic and chronic pollution of the environment by oil causes major ecological perturbations. Marine environments are especially vulnerable since oil spills of coastal regions and the open sea are poorly containable and mitigation is difficult. In addition to pollution through human activities, millions of tons of petroleum enter the marine environment every year from natural seepages. Despite its toxicity, a considerable fraction of petroleum oil entering marine systems is eliminated by the hydrocarbon-degrading activities of microbial communities, in particular by a remarkable recently discovered group of specialists, the so-called hydrocarbonoclastic bacteria (HCB).

Alcanivorax borkumensis, a paradigm of HCB and probably the most important global oil degrader, was the first to be subjected to a functional genomic analysis. This analysis has yielded important new insights into its capacity for (i) n-alkane degradation including metabolism, biosurfactant production and biofilm formation, (ii) scavenging of nutrients and cofactors in the oligotrophic marine environment, as well as (iii) coping with various habitat-specific stresses. The understanding thereby gained constitutes a significant advance in efforts towards the design of new knowledge-based strategies for the mitigation of ecological damage caused by oil pollution of marine habitats. HCB also have potential biotechnological applications in the areas of bioplastics and biocatalysis.

Metabolic Engineering and Biocatalytic Applications

The study of the fate of persistent organic chemicals in the environment has revealed a large reservoir of enzymatic reactions with a large potential in preparative organic synthesis, which has already been exploited for a number of oxygenases on pilot and even on industrial scale. Novel catalysts can be obtained from metagenomic libraries and DNA sequence based approaches.

Our increasing capabilities in adapting the catalysts to specific reactions and process requirements by rational and random mutagenesis broadens the scope for application in the fine chemical industry, but also in the field of biodegradation. In many cases, these catalysts need to be exploited in whole cell bioconversions or in fermentations, calling for system-wide approaches to understanding strain physiology and metabolism and rational approaches to the engineering of whole cells as they are increasingly put forward in the area of systems biotechnology and synthetic biology.

Bioplastic

Bioplastics or organic plastics are a form of plastics derived from renewable biomass sources, such as vegetable oil, corn starch, pea starch, or microbiota, rather than fossil-fuel plastics which are derived from petroleum. Some, but not all, bioplastics are designed to biodegrade.

Applications

Biodegradable bioplastics are used for disposable items, such as packaging and catering items (crockery, cutlery, pots, bowls, straws). Biodegradable bioplastics are also often used for organic waste bags, where they can be composted together with the food or green waste. Some trays and containers for fruit, vegetables, eggs and meat, bottles for soft drinks and dairy products and blister foils for fruit and vegetables are manufactured from bioplastics. Nondisposable applications include mobile phone casings, carpet fibres, and car interiors, fuel line and plastic pipe applications, and new electroactive bioplastics are being developed that can be used to carry electrical current. In these areas, the goal is not biodegradability, but to create items from sustainable resources.

Plastic Types

Starch based Plastics

Constituting about 50 percent of the bioplastics market, thermoplastic starch, such as Plastarch Material, currently represents the most important and widely used bioplastic. Pure starch possesses the characteristic of being able to absorb humidity, and is thus being used for the production of drug capsules in the pharmaceutical sector. Flexibiliser and plasticiser such as sorbitol and glycerine are added so the starch can also be processed thermo-plastically. By varying the amounts of these additives, the characteristic of the material can be tailored to specific needs (also called "thermo-plastical starch"). Simple starch plastic can be made at home shown by this method.

Cellulose based Plastics

Cellulose bioplastics are mainly the cellulose esters (cellulose acetate, nitrocellulose...) and their derivatives (celluloid...).

Some Aliphatic Polyesters

The aliphatic biopolyesters are mainly polyhydroxyalkanoates (PHAs) like the poly-3-hydroxybutyrate (PHB), polyhydroxyvalerate (PHV) and polyhydroxyhexanoate PHH.

Polylactic Acid (PLA) Plastics

Polylactic acid (PLA) is a transparent plastic produced from cane sugar or glucose. It not only resembles conventional petrochemical mass plastics (like PE or PP) in its characteristics, but it can also be processed easily on standard equipment that already exists for the production of conventional plastics. PLA and PLA blends generally come in the form of granulates with various properties, and are used in the plastic processing industry for the production of foil, moulds, tins, cups, bottles and other.

Poly-3-hydroxybutyrate (PHB)

The biopolymer poly-3-hydroxybutyrate (PHB) is a polyester produced by certain bacteria processing glucose or starch. Its characteristics are similar to those of the petroplastic polypropylene. The South American sugar industry, for example, has decided to expand PHB production to an industrial scale. PHB is distinguished primarily by its physical characteristics. It produces transparent film at a melting point higher than 130 degrees Celsius, and is biodegradable without residue.

Polyamide 11 (PA 11)

PA 11 is a biopolymer derived from natural oil. It is also known under the tradename Rilsan B, commercialized by Arkema. PA 11 belongs to the technical polymers family and is not biodegradable. Its properties are similar to those of PA 12, although emissions of greenhouse gases and consumption of nonrenewable resources are reduced during its production. Its thermal resistance is also superior to that of PA 12. It is used in high-performance applications like automotive fuel lines, pneumatic airbrake tubing, electrical cable antitermite sheathing, flexible oil and gas pipes, control fluid umbilicals, sports shoes, electronic device components, and catheters.

Bio-derived Polyethylene

The basic building block (monomer) of polyethylene is ethylene. This is just one small chemical step from ethanol, which can be produced by fermentation of agricultural feedstocks such as sugar cane or corn. Bio-derived polyethylene is chemically and physically identical to traditional polyethylene- it does not biodegrade but can be recycled. It can also considerably reduce greenhouse gas emissions.

Brazilian chemicals group Braskem claims that using its route from sugar cane ethanol to produce one tonne of polyethylene captures (removes from the environment) 2.5 tonnes of carbon dioxide while the traditional petrochemical route results in emissions of close to 3.5 tonnes. Braskem plans to introduce commercial quantities of its first bio-derived high density polyethylene, used in a packaging such as bottles and tubs, in 2010 and has developed a technology to produce bio-derived butene, required to make the linear low density polethylene types used in film production.

Genetically Modified Bioplastics

Genetic modification (GM) is also a challenge for the bioplastics industry. None of the currently available bioplastics- which can be considered first generation products- require the use of GM crops, although GM corn is the standard feedstock.

Looking further ahead, some of the second generation bioplastics manufacturing technologies under development employ the "plant factory" model, using genetically modified crops or genetically modified bacteria to optimise efficiency.

Environmental Impact

The production and use of bioplastics is generally regarded as a more sustainable activity when compared with plastic production from petroleum (petroplastic), because it relies less on fossil fuel as a carbon source and also introduces fewer, net-new greenhouse emissions if it biodegrades. They significantly reduce hazardous waste caused by oil-derived plastics, which remain solid for hundreds of years, and open a new era in packing technology and industry.

However, manufacturing of bioplastic materials is often still reliant upon petroleum as an energy and materials source. This comes in the form of energy required to power farm machinery and irrigate growing crops, to produce fertilisers and pesticides, to transport crops and crop products to processing plants, to process raw materials, and ultimately to produce the bioplastic, although renewable energy can be used to obtain petroleum independence.

Italian bioplastic manufacturer Novamont states in its own environmental audit that producing one kilogram of its starch-based product uses 500g of petroleum and consumes almost 80% of the energy required to produce a traditional polyethylene polymer. Environmental data from NatureWorks, the only commercial manufacturer of PLA (polylactic acid) bioplastic, says that making its plastic material delivers a fossil fuel saving of between 25 and 68 per cent compared with polyethylene, in part due to its purchasing of renewable energy certificates for its manufacturing plant.

A detailed study examining the process of manufacturing a number of common packaging items in several traditional plastics and polylactic acid carried out by Franklin Associates and published by the Athena Institute shows the bioplastic to be less environmentally damaging for some products, but more environmentally damaging for others. This study however does not consider the end-of-life of the products, thus ignores the possible methane emissions that can occur in landfill due to biodegradable plastics.

While production of most bioplastics results in reduced carbon dioxide emissions compared to traditional alternatives, there are some real concerns that the creation of a global bioeconomy could contribute to an accelerated rate of deforestation if not managed effectively. There are associated concerns over the impact on water supply and soil erosion.

Other studies showed that bioplastics represent a 42% reduction in carbon footprint.

On the other hand, bioplastic can be made from agricultural byproducts and also from used plastic bottles and other containers using microorganisms.

Bioplastics and Biodegradation

The terminology used in the bioplastics sector is sometimes misleading. Most in the industry use the term bioplastic to mean a plastic produced from a biological source. One of the oldest plastics, cellulose film, is made from wood cellulose. All (bio- and petroleum-based) plastics are technically biodegradable, meaning they can be degraded by microbes under suitable conditions. However many degrade at such slow rates as to be considered non-biodegradable.Some petrochemical-based plastics are considered biodegradable, and may be used as an additive to improve the performance of many commercial bioplastics. Non-biodegradable bioplastics are referred to as durable. The degree of biodegradation varies with temperature, polymer stability, and available oxygen content. Consequently, most bioplastics will only degrade in the tightly controlled conditions of industrial composting units. In compost piles or simply in the soil/water, most bioplastics will not degrade (e.g. PH), starch-based bioplastics will, however. An internationally agreed standard, EN13432, defines how quickly and to what extent a plastic must be degraded under commercial composting conditions for it to be called biodegradable. This is published by the International Organization for Standardization ISO and is recognised in many countries, including all of Europe, Japan and the US. However, it is designed only for the aggressive conditions of commercial composting units. There is no standard applicable to home composting conditions.

The term "biodegradable plastic" is often also used by producers of specially modified petrochemical-based plastics which appear to biodegrade. Traditional plastics such as polyethylene are degraded by ultra-violet (UV) light and oxygen. To prevent this process manufacturers add stabilising chemicals. However with the addition of a degradation initiator to the plastic, it is possible to achieve a controlled UV/oxidation disintegration process. This type of plastic may be referred to as *degradable plastic* or *oxy-degradable plastic* or *photodegradable plastic* because the process is not initiated by microbial action. While some degradable plastics manufacturers argue that degraded plastic residue will be attacked by microbes, these degradable materials do not meet the requirements of the EN13432 commercial composting standard.The bioplastics industry has widely criticised oxo-biodegradable plastics, which the industry association says do not meet its requirements.Oxo-biodegradable plastics- known as "oxos"- are conventional petroleum-based products with some additives that initiate degradation.ASTM standard used by oxo producers is just a guideline. It requires only 60% biodegradation, P-Life is an oxo plastic claiming biodegradability in soil at a temperature of 23 degrees Celsius reaches 66% after 545 days.Dr Baltus of the National Innovation Agency, said there was no proven evidence that bio-organisms are really able to consume and biodegrade oxo plastics.

Recycling

There are also concerns that bioplastics will damage existing recycling projects. Packaging such as HDPE milk bottles and PET water and soft drinks bottles is easily identified and hence setting up a recycling infrastructure has been quite successful in many parts of the world. However, plastics like PET do not mix with PLA, yielding unusable recycled PET if consumers fail to distinguish the two in their sorting. The problem could be overcome by ensuring distinctive bottle types or by investing in suitable sorting technology. However, the first route is unreliable, and the second costly.

Market

Because of the fragmentation in the market and still unsettled definitions it is difficult to estimate the total market size for bioplastics, but estimates put global consumption in 2006 at around 85,000 tonnes. In contrast, global consumption of all flexible packaging is estimated at around 12.3 million tonnes. COPA (Committee of Agricultural Organisation in the European Union) and COGEGA (General Committee for the Agricultural Cooperation in the European Union) have made an assessment of the potential of bioplastics in different sectors of the European economy:

Catering products: 450,000 tonnes per year

Organic waste bags: 100,000 tonnes per year

Biodegradable mulch foils: 130,000 tonnes per year

Biodegradable foils for diapers 80,000 tonnes per year

Diapers, 100% biodegradable: 240,000 tonnes per year

Foil packaging: 400,000 tonnes per year

Vegetable packaging: 400,000 tonnes per year

Tyre components: 200,000 tonnes per year.

Total 2,000,000 tonnes per year.

In the years 2000 to 2008, worldwide consumption of biodegradable plastics based on starch, sugar, and cellulose- so far the three most important raw materials- has increased by 600 %. The European Bioplastics trade group predicted annual capacity would more than triple to 1.5 million tons by 2011. BCC Research forecasts the global market for biodegradable polymers to grow at a compound average growth rate of more than 17 percent through 2012. Even so, bioplastics will encompass a small niche of the overall plastic market, which is forecast to reach 500 billion pounds (220 million tonnes) globally by 2010.

Cost

With the exception of cellulose, most bioplastic technology is relatively new and is currently not cost competitive with (petroplastics). Bioplastics do not yet reach the fossil fuel parity on fossil fuel-derived energy for their manufacturing, reducing the cost advantage over petroleum-based plastic. However, in certain, special applications bioplastics are already unbeatable because pure material costs form only a part of the entire product costs. For example, medical implants made of PLA, which dissolve in the body, save patients a second operation. Compostable mulch films for agriculture, already often produced from starch polymers, do not have to be collected after use and can be left on the fields.

Research and Development

- In the early 1950s, Amylomaize (>50% amylose content corn) was successfully bred and commercial bioplastics applications started to be explored.
- In 2004, NEC developed a flame retardant plastic, polylactic acid, without using toxic chemicals such as halogens and phosphorus compounds.
- In 2005, Fujitsu became one of the first technology companies to make personal computer cases from bioplastics, which are featured in their FMV-BIBLO NB80K line.
- In 2007 Braskem of Brazil announced it had developed a route to manufacture high density polyethylene (HDPE) using ethylene derived from sugar cane.
- In 2008, a University of Warwick team created a soap-free emulsion polymerization process which makes colloid particles of polymer dispersed in water, and in a one step process adds nanometre sized silica-based particles to the mix. The newly developed technology might be most applicable to multi-layered biodegradable packaging, which could gain more robustness and water barrier characteristics through the addition of a nano-particle coating.

Certification

Biodegradability- EN 13432, ASTM D6400

The EN 13432 industrial standard is arguably the most international in scope and compliance with this standard is required to claim that a product is compostable in the European marketplace. In summary, it requires biodegradation of 90% of the materials in a commercial composting unit within 180 days. The ASTM 6400 standard is the regulatory framework for the United States and sets a less stringent threshold of 60% biodegradation within 180 days, again within commercial composting conditions.

The "compostable" marking found on many items of packaging indicates that the package complies with either of the two standards mentioned above. However, the marking is not owned by either regulatory body but by third party trade associations representing companies making or selling biodegradable plastics. In Europe, this is European Bioplastics, in the U.S. it is the Biodegradable Products Institute.

Many starch based plastics, PLA based plastics and certain aliphatic-aromatic co-polyester compounds such as succinates and adipates, have obtained these certificates. Additivated plastics sold as photodegradable or Oxo Biodegradable do not comply with these standards in their current form.

Biobased- ASTM D6866

The ASTM D6866 method has been developed to certify the biologically derived content of bioplastics. Cosmic rays colliding with the atmosphere mean that some of the carbon is the radioactive isotope carbon-14. CO_2 from the atmosphere is used by plants in photosynthesis, so new plant material will contain both carbon-14 and carbon-12. Under the right conditions, and over geological timescales, the remains of living organisms can be transformed into fossil fuels. After ~100,000 years all the carbon-14 present in the original organic material will have undergone radioactive decay leaving only carbon-12. A product made from biomass will have a relatively high level of carbon-14, while a product made from petrochemicals will have no carbon-14.The percentage of renewable carbon in a material (solid or liquid) can be measured with an accelerator mass spectrometer.

There is an important difference between biodegradability and biobased content. A bioplastic such as high density polyethylene (HDPE) can be 100% biobased (i.e. contain 100% renewable carbon), yet be non-biodegradable. These bioplastics such HDPE play nonetheless an important role in greenhouse gas abatement, particularly when they are combusted for energy production. The biobased component of these bioplastics is considered carbon-neutral since their origin is from biomass.

Cyanobacteria

Cyanobacteria (also known as blue-green algae, blue-green bacteria, and Cyanophyta) is a phylum of bacteria that obtain their energy through photosynthesis. The name "cyanobacteria" comes from the colour of the bacteria. The ability of cyanobacteria to perform oxygenic photosynthesis is thought to have converted the early reducing atmosphere into an oxidizing one, which dramatically changed the composition of life forms on Earth by stimulating biodiversity and leading to the near-extinction of oxygen-intolerant organisms. According to endosymbiotic theory, chloroplasts in plants and eukaryotic algae have evolved from cyanobacteria via endosymbiosis.

Ecology

Cyanobacteria can be found in almost every conceivable environment, from oceans to fresh water to bare rock to soil. They can occur as planktonic cells or form phototrophic biofilms in fresh water and marine environments, they occur in damp soil, or even temporarily moistened rocks in deserts. A few are endosymbionts in lichens, plants, various protists, or sponges and provide energy for the host. Some live in the fur of sloths, providing a form of camouflage. Aquatic cyanobacteria are probably best known for the extensive and highly visible blooms that can form in both freshwater and the marine environment and can have the appearance of blue-green paint or scum. The association of toxicity with such blooms has frequently led to the closure of recreational waters when blooms are observed. Marine bacteriophage are a significant parasite of unicellular, marine cyanobacteria. When they infect cells they lyse them releasing more phages into the water.

Characteristics

Cyanobacteria include unicellular and colonial species. Colonies may form filaments, sheets or even hollow balls. Some filamentous colonies show the ability to differentiate into several different cell types: vegetative cells, the normal, photosynthetic cells that are formed under favourable growing conditions; akinetes, the climate-resistant spores that may form when environmental conditions become harsh; and thick-walled heterocysts, which contain the enzyme nitrogenase, vital for nitrogen fixation. Heterocysts may also form under the appropriate environmental conditions (anoxic) when fixed nitrogen is scarce. Heterocyst-forming species are specialized for nitrogen fixation and are able to fix nitrogen gas into ammonia (NH_3), nitrites (NO"2) or nitrates (NO"3) which can be absorbed by plants and converted to protein and nucleic acids [atmospheric nitrogen cannot be used by plants directly]. Rice crops utilize healthy populations of nitrogen-fixing cyanobacteria in some rice paddy fertilizers.

Many cyanobacteria also form motile filaments, called hormogonia, that travel away from the main biomass to bud and form new colonies elsewhere. The cells in a hormogonium are often thinner than in the vegetative state, and the cells on either end of the motile chain may be tapered. In order to break away from the parent colony, a hormogonium often must tear apart a weaker cell in a filament, called a necridium.

Each individual cell of a cyanobacterium typically has a thick, gelatinous cell wall. They differ from other gram-negative bacteria in that the quorum sensing molecules autoinducer-2 and acyl-homoserine lactones are absent. They lack flagella, but hormogonia and some species may move about by gliding along surfaces. Many of the multi-cellular filamentous forms of *Oscillatoria* are capable of a waving motion; the filament oscillates back and forth. In water columns some cyanobacteria float by forming gas vesicles, like in archaea. These vesicles are not organelles as such. They are not bounded by lipid membranes but by a protein sheath.

Some of these organisms contribute significantly to global ecology and the oxygen cycle. The tiny marine cyanobacterium *Prochlorococcus* was discovered in 1986 and accounts for more than half of the photosynthesis of the open ocean. Many cyanobacteria even display the circadian rhythms that were once thought to exist only in eukaryotic cells.

Photosynthesis

Cyanobacteria account for 20–30% of Earth's photosynthetic productivity and convert solar energy into biomass-stored chemical energy at the rate of ~450 TW. Cyanobacteria utilize the energy of sunlight to drive photosynthesis, a process where the energy of light is used to split water molecules into oxygen, protons, and electrons. While most of the high-energy electrons derived from water are utilized by the cyanobacterial cells for their own needs, a fraction of these electrons are donated to the external environment via electrogenic activity. Cyanobacterial electrogenic activity is an important microbiological conduit of solar energy into the biosphere.

Cyanobacteria have an elaborate and highly organized system of internal membranes which function in photosynthesis. Cyanobacteria get their name from the bluish pigment phycocyanin, which they use to capture light for photosynthesis. Photosynthesis in cyanobacteria generally uses water as an electron donor and produces oxygen as a by-product, though some may also use hydrogen sulfide as occurs among other photosynthetic bacteria. Carbon dioxide is reduced to form carbohydrates via the Calvin cycle. In most forms the photosynthetic machinery is embedded into folds of the cell membrane, called thylakoids. The large amounts of oxygen in the atmosphere are considered to have been first created by the activities of ancient cyanobacteria. Due to their ability to fix nitrogen in aerobic conditions they are often found as symbionts with a number of other groups of organisms such as fungi (lichens), corals, pteridophytes (*Azolla*), angiosperms (*Gunnera*) etc.

Many cyanobacteria are able to reduce nitrogen and carbon dioxide under aerobic conditions, a fact that may be responsible for their evolutionary and ecological success. The water-oxidizing photosynthesis is accomplished by coupling the activity of photosystem (PS) II and I (Z-scheme). In anaerobic conditions, they are also able to use only PS I — cyclic photophosphorylation — with electron donors other than water (hydrogen sulfide, thiosulphate, or even molecular hydrogen) just like purple photosynthetic bacteria. Furthermore, they share an archaeal property, the ability to reduce elemental sulfur by anaerobic respiration in the dark. Their photosynthetic electron transport shares the same compartment as the components of respiratory electron transport. Their plasma membrane contains only components of the respiratory chain, while the thylakoid membrane hosts both respiratory and photosynthetic electron transport.

Attached to thylakoid membrane, phycobilisomes act as light harvesting antennae for the photosystems. The phycobilisome components (phycobiliproteins) are responsible for the blue-green pigmentation of most cyanobacteria. The variations to this theme is mainly due to carotenoids and phycoerythrins which give the cells the red-brownish coloration. In some cyanobacteria, the colour of light influences the composition of phycobilisomes. In green light, the cells accumulate more phycoerythrin, whereas in red light they produce more phycocyanin. Thus the bacteria appear green in red light and red in green light. This process is known as complementary chromatic adaptation and is a way for the cells to maximize the use of available light for photosynthesis.

A few genera, however, lack phycobilisomes and have chlorophyll *b* instead (*Prochloron, Prochlorococcus, Prochlorothrix*). These were originally grouped together as the prochlorophytes or chloroxybacteria, but appear to have developed in several different lines of cyanobacteria. For this reason they are now considered as part of the cyanobacterial group.

Relationship to Chloroplasts

Chloroplasts found in eukaryotes (algae and plants) likely evolved from an endosymbiotic relation with cyanobacteria. This endosymbiotic theory is supported by various structural and genetic similarities Primary chloroplasts are found among the "true plants" or green plants — species ranging from sea lettuce to evergreens and flowers which contain chlorophyll *b* — as well as among the red algae and glaucophytes, marine species which contain phycobilins. It now appears that these chloroplasts probably had a single origin, in an ancestor of the clade called Primoplantae. Other algae likely took their chloroplasts from these forms by secondary endosymbiosis or ingestion.

It was once thought that the mitochondria in eukaryotes also developed from an endosymbiotic relationship with cyanobacteria; however, it is now suspected that this evolutionary event occurred when aerobic bacteria were engulfed by anaerobic host cells. Mitochondria are believed to have originated not from cyanobacteria but from an ancestor of *Rickettsia*.

Relationship to Earth History

Stromatolites of fossilized oxygen-producing cyanobacteria have been found from 2.8 billion years ago, possibly as old as 3.5 billion years ago.

The biochemical capacity to use water as the source for electrons in photosynthesis evolved once, in a common ancestor of extant cyanobacteria. The geologic record indicates that this transforming event took place early in our planet's history, at least 2450-2320 million years ago (mya), and probably much earlier. Geobiological interpretation of Archean (>2500 mya) sedimentary rocks remains a challenge; available evidence indicates that life existed 3500 mya, but the question of when oxygenic photosynthesis evolved continues to engender debate and research.

A clear paleontological window on cyanobacterial evolution opened about 2000 mya, revealing an already diverse biota of blue-greens. Cyanobacteria remained principal primary producers throughout the Proterozoic (2500-543 mya), in part because the redox structure of the oceans favored photoautotrophs capable of nitrogen fixation.

The most common cyanobacterial structures in the fossil record are the mound-producing stromatolites and related oncolites. Indeed, these fossil colonies are so common that paleobiology, micropaleontology and paleobotany cite the Pre-Cambrian and Cambrian period as an "age of stromatolites" and an "age of algae."

Green algae joined the blue-greens as major primary producers on continental shelves near the end of the Proterozoic, but only with the Mesozoic era (251-65 mya) radiations of dinoflagellates, coccolithophorids, and diatoms did primary production in marine shelf waters take modern form.

Today, the blue-green bacteria remain critical to marine ecosystems as primary producers in oceanic gyres, as agents of biological nitrogen fixation, and—in modified form—as the plastids of marine algae.

Bibliography

Alfred Byrd Graf: *Advances in Plant Physiology*, Rajat Pub, Delhi, 2008.

Altman, Arie, : : *Agricultural Biotechnology*, New York: Marcel Dekker, 1998.

Ashraf, Syed Mahboob: *Handbook of Fruit and Vegetable Products*, Agrobios, Delhi, 2008.

Babar, Md. : *Hydrogeomorphology : Fundamentals Applications and Techniques*, New India Pub, Delhi, 2005.

Balch C.N.C, P. : *Prescription for Nutritional Healing*, New York, U.S.A: Avery Publishing Group, 1997.

Baligar, V C and R B Clark: *Physiology of Crop Production*, International Book, Delhi, 2007.

Bandyopadhyay, P C : *Breeding and Crop Production*, Gene Tech Books, Delhi, 2007.

Bhutani, R.C.: *Fruit and Vegetable Preservation*, Biotech Books, Delhi, 2003.

Boyer , J.S.: *Measuring the Water Status of Plants and Soils*, Academic Press, N.Y., 1995.

Bushnell, R.B. : *Dry Cow Feeding and Management*, A Western Regional Extension Publication, 1979.

Chadha K. L. and Pareek O. P.: *Advances in Horticulture: Fruit Crops*, New Delhi, Malhotra Publishing House, 1993.

Chaudhary, Vikas : *Entomology and Pest Management*, Navyug, Delhi, 2008.

Collymore L.: *Fruit Production in Barbados*, Port of Spain, Trinidad and Tobago, 1996.

Coste R.: *Coffee: the Plant and the Product*, London, MacMillan, 1992.

Cristobal Noe Aguilar : *Food Science and Food Biotechnology in Developing Countries*, Asiatech Pub, 2008.

Cronquist, A.: *The Evolution and Classification of Flowering Plants*, New York Botanical Garden, Bronx, New York., 1988

Dharamvir Hota: *Modern Biotechnology in Plant Breeding,* Gene Tech Books, Delhi, 2007.

Doijode S. D.: *Seed Germination in Fruits,* New Delhi, Malhotra Publishers, 1993.

Ferentinos L.: *Proceeding of the Sustainable Taro Culture for the Pacific Conference,* Honolulu, HITAHR, 1993.

Grace, Eric S. : *Biotechnology Unzipped: Promises and Realities,* Washington, D.C.: Joseph Henry Press, 1997.

Gupta, S.K.: *Disease Problems in Vegetable Production,* Scientific, Delhi, 2006.

Herminie Broedel Kitchen: *Soils and Crops : Diagnostic Techniques,* Satish Serial Publishing, Allahabad, 2004.

Hota, Dharamvir : *Modern Biotechnology in Plant Breeding,* Gene Tech Books, Delhi, 2007.

Jacobson, M. : *Safe Food: Eating Wisely in a Risky World,* Washington, DC: Living Planet Press, 1991.

James S Shoemaker: *Culture of Vegetable Growing,* Asiatic Pub, Delhi, 2006.

Johnson-Green, Perry : *Introduction to Food Biotechnology,* Boca Raton, FL: CRC Press, 2002.

Kumar, N. : *Breeding of Horticultural Crops : Principles and Practices,* New India Pub, Delhi, 2006.

Lata Bhattacharya: *Biochemistry of Nutrition,* Discovery, Delhi, 2010.

Macself, A.J. : *Soils and Fertilizers,* Satish Serial Pub, Delhi, 2005.

Narang, R.K.: *Fruit and Vegetable Preservation Techniques,* APH Pub, Delhi, 2010.

Nobel, P. S.: *Physicochemical and Environmental Plant Physiology,* Academic Press, San Diego, 1999.

Oldham P.: *Cost of Production of Major Tree Crops in Dominica,* Roseau, Ministry of Agriculture, 1991.

Pareek O. P.: *Advances in Horticulture: Fruit Crops,* New Delhi, Malhotra Publishing House, 1993.

Parmar, B.S. : *Pesticides, Crop Protection and Environment,* Oxford & IBH, Delhi, 1995.

Patel, IS and GM Patel: *Botanical Pesticides for Pest Management,* Scientific, Delhi, 2008.

Paul B. : *Food Biotechnology in Ethical Perspective,* Aspen, CO: Aspen Publishers, 1997.

Politycka, B and C L Goswami: *Plant Physiology : Research Methods,* Scientific, Delhi, 2007.

Purohit, S.S. : *Organic Farming : Biocontrol and Biopesticide Technology*, Agrobios, Delhi, 2008.

Ramann, E. : *The Evolution and Classification of Soils*, Asiatic Pub, Calcutta, 2006.

Ranganathan, M.S. and Sriram Sridhar : *Enzyme and Food Biotechnology*, Wisdom Press, 2011.

Saini, M.L. : *Plant Breeding and Crop Improvement*, CBS, Delhi, 1997.

Selden, G. : *Body Electric: Electromagnetism and the Foundation of Life*, New York, U.S.A: William Morrow & Co., 1987.

Sharma, R.C. : *Diseases of Horticultural Crops: Fruits*, Indus, Delhi, 1999.

Sharma, Rajni : *An Introduction to Plant Morphology*, Campus, Delhi, 2004.

Srivastava, Seema : *Plant Physiology and Biochemistry*, Campus Books, Delhi, 2009.

Thomas Bernauer : *Genes, Trade and Regulation : The Seeds of Conflict in Food Biotechnology*, Universities Press, 2005.

Tomar, S.K. and Gunjan Goel : *Applied Dairy and Food Microbiology*, Agrotech, 2005.

Tripathi, Pramila: *Botanical Pesticides in the Management of Post-harvest Fruit Diseases*, Daya, Delhi, 2005.

Tripathy, S.N. : *Food Biotechnology*, Dominant, 2004.

Vanangamudi, K. : *Principles and Methods of Plant Breeding*, International Book, Delhi, 2005.

Yadav, Keshav Prasad : *Application of Morphometry in Geomorphology*, Radha Pub, Delhi, 2008.

Yoshida T.: *Cultivation of Citrus Genetic Resources for Evaluation of Characteristics*, Tokyo, JICA, 1996.

Index

A

B

□□□